Effect of Fly Ash
on the Physical Properties
of Illite-Based Ceramics

Authors:
Tomáš Húlan
Ján Ondruška

Reviewers:
Igor Medveď
Zbigniew Suchorab

Published by **Materials Research Forum LLC**
Millersville, PA 17551, USA

Published as part of the book series
Materials Research Foundations
Volume 130 (2022)
ISSN 2471-8890 (Print)
ISSN 2471-8904 (Online)

Print ISBN 978-1-64490-206-6
ePDF ISBN 978-1-64490-207-3

Distributed worldwide by

Materials Research Forum LLC
105 Springdale Lane
Millersville, PA 17551
USA
http://www.mrforum.com

Printed in the United States of America
10 9 8 7 6 5 4 3 2 1

Table of Contents

Introduction

In the past, the basic building materials were wood, stone, or iron. Later, people discovered the possibility of using dried clay shaped into bricks to build houses. At first, the bricks were merely allowed to dry in the sun to obtain the desired properties. Then it was found out that if such a brick was fired at a high temperature, around 700 °C, some of its properties would improve, especially durability and strength. People thus produced an artificial material that they could use to their great advantage in constructing their dwellings. In addition to bricks, the group of ceramic materials used in construction includes roof tiles, tiles, chimney liners, ceiling slabs, or parts of drains.

The primary way to reduce the cost of production of building materials is through research into the use of cheaper basic raw materials while maintaining the same quality and parameters of the final product. For example, manufacturers add various additives to ceramic products to improve their final properties. In addition, the production process must be taken into account, as the raw unfired ceramic shard must meet specific requirements for strength, hardness, water absorption, etc.

Closely linked to cost reduction is the great pressure on electricity production. Currently, many countries are moving away from nuclear power generation (especially because of the recent Fukushima accident in Japan in 2011). This shortfall has to be replaced somehow, which is currently only possible with coal-combusting thermal powerplants. This translates into an increased production of fly ash, the solid residue left over from coal combustion. In a 1 000 MW powerplant with an average consumption of 12 000 t of lignite per day, fly ash production can be as high as 2 000 t per day [1]. In Slovakia, waste from energy production in power and heating plants accounts for 70% of the total waste production [2]. If fly ash were released into the air, it would represent a significant ecological and environmental burden for the wider environment. Therefore, fly ash must be separated from the flue gas by mechanical or electrostatic precipitators and subsequently stored. However, storage is also not an ecologically ideal solution, mainly due to the increased toxicity of fly ash [3]. The costs associated with ash storage are reflected in the price of electricity. On the other hand, fly ash represents a potential source of a secondary raw material. Therefore, there is a need to find a suitable use for fly ash as a secondary raw material, which could bring the following economic and environmental benefits:

- Reducing the environmental burden from ash storage;
- Reduction of storage costs by the ash producer (and thus reduction of the energy price);
- Reduced product production costs by the ash buyer, plus saving of other natural resources that the ash will replace (which will again be reflected in a lower price for the consumer).

One of the most common methods of fly ash processing is its use as an additive in the production of concrete [4]. The use of fly ash in agriculture, as a fertilizer or as an additive to reduce soil acidity, is also well known [1]. More or less successful attempts to produce glass ceramics and geopolymers from fly ash can also be found in the literature [5, 6]. An overview of research dealing with fly ash utilization outside the construction industry can be found, e.g., in [7]. For several years, researchers have also been trying to find applications of fly ash in the production of ceramic materials [1–18]. The studies that have dealt with the use of fly ash from power plants that do not burn lignite but, for example, municipal waste, will not be addressed in this work.

However, a large part of these attempts remains only at the level of academic research. The reason for this is the hesitation on the part of producers due to the ambiguous results of individual studies. The variability of fly ash composition over time appears to be a major problem. Even greater variability can be expected in the case of fly ash from different thermal power plants. This is why it is crucial to investigate the possibilities of utilizing fly ash from different sources individually.

The content of this monograph is to investigate the possibilities of using fly ash in the production of building ceramics. Specifically, we will focus on three types of fly ash from a thermal power plant which used two lignite combustion technologies – pulverized-granulation combustion and fluidized bed combustion. In the ceramics industry, a grog is used in the production of traditional ceramics to prevent rapid volume changes. In this work, a procedure was chosen in which a part of the traditional grog was gradually replaced by three types of power plant fly ash. Subsequently, the influence of this substitution on the resulting properties of the ceramic body as well as on the processes occurring in it during heat treatment was monitored.

Clay rock with a majority of the mineral illite was chosen as a plastic component for the production of ceramic bodies. Illite is a mineral frequently found in clay materials used in the ceramic industry [18]. Its high abundance is due to its relatively high stability under normal conditions [19]. Due to the properties mentioned above, it was used as a primary raw material for preparing the samples. Since it is almost a single-component material, it will be easier to form analyses and conclusions. Since illite is a common component of raw materials for the production of traditional ceramics, it is anticipated that the conclusions drawn will be applicable to illitic soils from other locations.

In this monograph, the aim was to use as many relevant experimental methods as possible, but mainly to look for connections between the results of the individual analyses. The material parameters of the experimental mixtures are discussed concerning their use in the production of building ceramics. In addition, a link between changes in material parameters and microstructural evolution is sought, whether based on knowledge from the literature, analysis results, or hypotheses. Many of the analyses, such as X-ray diffraction, scanning electron microscopy, and others, have been carried out in collaboration with the departments of Tallinn University of Technology, Czech Technical University in Prague, Charles University in Prague, and Slovak University of Technology in Bratislava.

1. Ceramics

The origin of the term "ceramics" is taken to mean the Greek word keramos which originally referred to a drinking horn and later to other vessels serving this purpose [20]. At present, the name ceramics is used for shaped ceramic products, fired ceramic, or the whole industry [21]. Ceramics is the oldest human-made artificial material. The first known ceramic products date back to the 7th century BC [20].

In terms of structure, ceramics is defined as an inorganic non-metallic substance with a heterogeneous polycrystalline structure prepared by sintering at high temperatures [20]. It consists of crystalline substances of different compositions and arrangements, glass phase, and pores [21]. The manufacturing process of the ceramic product begins with shaping the powders of the base material after mixing with water or pressing the dry powder, proceeds with drying, and ends with firing. By firing, the ceramic body is compacted, strengthened, and obtains the required performance properties such as strength, hardness, durability, chemical resistance, etc. As opposed to the metallurgical or glass technology principle, firing temperature is generally lower than the melting temperature [22]. Ceramics are characterized by high chemical resistance, durability, abrasion resistance, compressive strength, and brittleness [22]. Another characteristic feature of ceramic materials is a relatively low reproducibility of the measurement results of their properties. This is due to the presence of defects in the form of pores, agglomerates, cracks, or residual stresses, the most dangerous being cracks [23].

In the past, ceramic research and production were based on empirical procedures. In addition to the lengthy nature, this approach caused problems in the event of a change in raw materials or production process. Later, in an effort to overcome this problem, the properties of the individual isolated phases of the ceramics, their formation and arrangement in the ceramic mass were investigated. Such a process has proven to be an essential part of physical and chemical ceramic research even today, although empirical processes have not been completely abandoned due to the complicated structure and phase composition of traditional ceramics in particular [21].

1.1 Classification of ceramics

Ceramic materials can be divided according to different aspects [20, 22, 23]:

1. According to structure:
   a. *Fine ceramics*. The individual particles have a diameter of about 0.1 mm and cannot be recognized by the naked eye. The structure of the shards is observable only by microscopy, and the pore content is small. These include, for example, tiles, wall tiles, sanitaryware and pottery, dental ceramics, etc.

     b. *Coarse ceramics* has a clear proportion of particles with a diameter of 0.1 mm to 5 mm. However, a certain proportion of particles below 2 μm is also required for shaping. This includes brick products, sewer, and chemical stoneware, garden ceramics, refractory products, etc.

2. According to the content of chemical compounds:

     a. *Silicate ceramics.* These are mainly products of building and consumer ceramics (bricks, tiles, containers, etc.), but also electro-ceramics and refractory products.

     b. *Oxide ceramics.* These include primary refractory products, sintered oxides, corundum, etc.

     c. *Non-oxide ceramics.* These include carbides, borides, nitrides, and cermets.

3. According to usage:

     a. *Traditional ceramics.* Ceramic materials are made from common, naturally occurring raw materials such as clay minerals or quartz sand.

       i. *Building ceramics.* These are products permanently built into the building, such as bricks, lintels, roof tiles, wall tiles, or floor tiles.

       ii. *Refractory ceramics.* It is used for high-temperature applications and includes fireclay, dinas, etc.

       iii. *Ceramics for industrial purposes.* These include, for example, grinding and cutting tools, grinding bodies, insulators, components for electrical engineering, and others.

       iv. *Consumer ceramics* (containers, porcelain, sanitary ware, art ceramics, etc.).

     b. *Progressive ceramics.* Ceramics made from very pure synthetic raw materials (Al_2O_3, SiO_2, SiC, etc.). The products have unique properties for special applications.

       i. *Functional ceramics.* It has a dynamic character, which means changing its properties under the influence of external forces or fields. This includes materials with magnetic, dielectric, piezo, ferro, and pyroelectric, resistive, semiconductor, and superconducting properties. They create parts of integrated circuits, transistors, sensors, etc.

       ii. *Structural ceramics.* Its passive character is used, i.e., the stability of structure and properties under the effect of mechanical and thermal load or chemically aggressive environment.

       iii. *Bioceramics.* Used in healthcare (artificial joints, prostheses, teeth, etc.).

1.2   Structure and properties of ceramics

As with any material, the macroscopic properties of the ceramics result from its chemical and phase composition, the nature of the bonds at the atomic level, and the stability of the real structure of the individual phases and their interfaces. The predominant type of chemical bonds in ceramics is the ionic or covalent bond. Compared to metals, ceramics have a more complex structure. The nature of the links significantly limits the number of slip planes. This limits the movement of dislocations, which is one of the causes of the fragile nature of the ceramic. Microstructure (spatial arrangement of particles, including pores and cracks), bond type, and imperfect bonding at grain boundaries cause low plasticity of the ceramics and lead to brittle, mostly intergranular fracture [23].

Building ceramics are chemically classified as silicate. The basic bond in silicate ceramic is the bond of silicon with oxygen. The Si-O binding energy is 368 kJ mol^{-1}. This relatively high binding energy implies considerable stability of the $(SiO_4)^{4-}$ complex under normal conditions [3]. Regular arrangement of the crystalline lattice occurs only at small distances. When the crystals grow from the melt, the individual crystals interfere with their growth, resulting in a compact polycrystalline substance. The crystal lattice contains a number of failures that affect reactivity. As the number of disorders increases, the substance passes from the crystalline to the amorphous form. Such a structure is typical of the glass state [22].

In terms of the content of the individual components, the ceramics can be considered as a composite material. Traditional ceramics can be characterized as a system [ceramics – glass – pores]. The individual particles form agglomerates which are embedded in a glass matrix at sufficiently high firing temperatures. The sizes and shapes of the particles influence the properties of such a system. Their anisotropic properties can then be influenced by their orientation [24]. An important factor affecting the physical properties is a good contact of the individual phases in the contact zone. It is formed when the melted (low viscosity) glass phase is capable of wetting the solid phase and subsequently solidifying without the stress caused by the thermal contraction breaking the joint [22].

The phase composition of ceramic materials predetermines their properties. In traditional ceramics, the amorphous phase (glass) is often the dominant component. It is based on SiO_2 and Al_2O_3 crosslinking oxides [25]. In the viscous flow sintering process, an important parameter of the glass phase is its viscosity. Low viscosity facilitates sintering by viscous flow. Alkaline earth metal oxides (Ca, Mg) slightly decrease the viscosity of the glass phase. Alkali metal oxides (Na, K) strongly modify the amorphous phase network and significantly reduce its viscosity [26]. As the temperature rises, the glass gradually moves from the solid phase to the low viscosity phase. This transition does not occur at a narrowly defined temperature range, as is the case with the melting of the crystalline phase. However, there is the glass transition temperature (T_g) above which the viscosity of the glass begins to drop sharply as the temperature rises. With the reduction in the viscosity of the glass by various impurities, T_g is often also shifted to lower temperatures, which is advantageous for the production economy. For example, to increase the formation of a low viscosity glass phase, a potassium feldspars are often added to traditional ceramic compositions. However, the risk of deformations of the product in the firing process also increases as the proportion of low viscosity glass is increased.

In addition to the amorphous phase and pores, ceramic materials also contain a crystalline phase – minerals. The amount of crystalline phase depends on the conditions that will allow their formation. It is essential that a suitable stoichiometric ratio of chemical elements is created in a particular area of the material. Moreover, a sufficient mobility of the atoms must be ensured so that they can occupy proper places in the crystal lattice. This is made possible mainly by the presence of a liquid phase (melt), which in turn is conditional on a sufficiently high temperature for the melt to be formed. Another important factor is the number of nuclei from which crystals grow. In traditional ceramics, heterogeneous crystallization is the most common, i.e., crystallization on foreign impurities [25]. Some oxides such as Fe_2O_3, TiO_2, and ZrO_2 are considered to be the so-called nucleating agents (or nucleants) that facilitate crystallization [25].

The various minerals, by their properties, influence the macroscopic properties of the ceramic body. For example, mullite ($3Al_2O_3 \cdot 2SiO_2$), which crystallizes in the form of long needle-like crystals, increases the strength of ceramic materials [26]. Quartz (SiO_2) reduces firing shrinkage but is dangerous in cooling, as it shrinks and modifies cracks in its surroundings [21]. Corundum (Al_2O_3) has a high hardness, strength, low electrical conductivity, heat resistance, and texture stability. Anorthite ($2SiO_2 \cdot Al_2O_3 \cdot CaO$) increases strength, reduces firing shrinkage, and increases chemical resistance [20, 27]. Cordierite improves resistance to thermal shocks due to its small coefficient of thermal expansion [28].

1.3   Raw materials of the traditional ceramics

The production of traditional ceramics is based on the use of natural raw materials that have the ability to form a well-shapeable plastic mass when mixed with water. The body shaped from such a mass retains its shape and does not deform itself by its own weight. During firing, the product is strengthened and fixed in shape. The raw materials used in the manufacture of the ceramic body meeting these requirements are called the plastic component. Natural raw materials of this sort are clays, such as kaolins and ball clays [21].

The more plastic raw material is used during shaping, the more shrinkage occurs during drying and firing. For this reason, substances reducing these side effects are added to the ceramic mixture. These substances do not form a plastic mass with water and are therefore referred to as non-plastic components. They are divided into tempers, fluxes, and pore formers [22].

The tempers are mostly non-porous, volumetrically and structurally stable during shaping and firing. Their chemical composition is close to that of the raw mixture. They reduce the formability of the plastic mass and are usually coarser than plastic components. When fired, they react with the other components of the mixture to form a dense shard. Typical tempers are quartz sand, corundum, chamotte, or various industrial wastes (e.g., from ceramic production) [21, 22].

The fluxes behave similarly to the tempers during shaping and drying. However, they aid in the formation of melt during firing, reducing the viscosity of the glass phase and the sintering temperature. Typical fluxes are feldspars (potassium, sodium, and calcium). Other fluxes used are raw materials containing oxides of CaO, MgO, Fe_2O_3, K_2O, Na_2O, etc. These oxides form low-melting compounds with SiO_2 [21, 22]. Recently, attention has been paid to the mineral wollastonite ($CaSiO_3$), which is a source of CaO without releasing gases that would increase porosity. CaO reduces the viscosity of the glass phase at high temperatures and thus helps to compact the ceramic shard. At the same time, firing shrinkage and mechanical strength are reduced [29, 30].

The pore former is a non-plastic component that contains pores or burns out during firing. This creates pores in the ceramic body, which reduces the bulk density of the product after firing. The disadvantage is that the ceramic body thus loses its strength. On the other hand, the advantage is that its thermal conductivity is reduced. Moreover, by producing heat in the body volume when the lighter burns out, it is possible to achieve the energy savings necessary for firing the product. For example, sawdust, paper waste, coal dust and others [21, 22] are used as pore formers.

This traditional division is meaningless in the case of new ceramic materials in which organic substances (plasticizers) are often used as carriers of formability, or in which other forming processes (e.g., pressing) are used [21].

The first topic of this work is the replacement of traditional tempers by power plant fly ash and the monitoring of the effect of this substitution on the physical properties of building ceramics. However, due to the content of low-melting oxides of CaO, Fe_2O_3, Na_2O, and others, fly ash can also partially act as a flux and lower the sintering temperature [1]. In addition, the fly ash may also contain unburned coal residues or porous particles, so that it may also appear to be a pore former in this sense [10, 16].

1.4   Firing of ceramics

Firing is a thermal process in which the ceramic material obtains the desired properties. It consists of several stages, the most important of which is sintering. Even before the high-temperature sintering phase, different processes occur in the clay-based ceramic material. These include the release of physically bound water, thermal decomposition of clay minerals, heating of organic residues, and more. High-temperature processes in silicate ceramics are described by solid-phase reactions, melting processes, glass formation, and crystallization [22]. Generally, these processes are called phase changes. They do not occur in the entire system at the same time but begin in small areas (nuclei) from which they then spread further.

1.4.1 Typical processes occurring during the firing of ceramics

Table 1.1 – Common reactions in ceramic shred during heating
(end – endothermic reaction, exo – exothermic reaction, m – weight loss occurs).

Temperature [°C]	Reaction	Note
90 – 200	Removal of physically bound water	[22], end, m
120 – 170	Decomposition of calcium sulphate dihydrate: $CaSO_4 \cdot 2H_2O \rightarrow CaSO_4 \cdot 1/2H_2O$	[31], end, m
180 – 220	Decomposition of calcium sulphate hemihydrate: $CaSO_4 \cdot 1/2H_2O \rightarrow CaSO_4$	[31], end, m
210 – 240	Dehydroxylation of gibbsite with possible formation of boehmite at (260 – 300) °C	[32], end, m
250 – 400	Thermooxidation of organic compounds	exo, m
470 – 520	Decomposition of boehmite	[32], end, m
450 – 550	Oxidation of pyrite	[33], exo, m
450 – 700	Dehydroxylation of illite (two-step)	[34], end, m
500 – 650	Dehydroxylation of kaolinite	[22], end, m
~450	Decomposition of $Ca(OH)_2$ (portlandite)	[35], end, m
573	Modification transition of quartz	[22], end
620 – 760	Dehydroxylation of montmorillonite	[22], end, m
750 – 820	Decomposition of $CaCO_3$ (calcite) and $CaMg(CO_3)_2$ (dolomite)	[36], end, m
800 – 900	Formation of Al-Si spinel structure from kaolinite	[22], exo
900 – 1 000	Formation of glassy phase from illite	[34], end
950 – 1 150	Decomposition of $CaSO_4$	[37], end, m
1 100 – 1 200	Formation of anorthite	[22], exo
1 100 – 1 300	Formation of mullite	[38], exo
1 200 – 1 300	Formation of cordierite	[22], exo

Some typical processes that occur during the heating of the ceramic material are summarized in Table 1.1. The temperature intervals are only indicative. Their beginnings and ends may shift, depend on the degree of crystallization, impurities, granulometry, effluent efficiency of decomposition products, heating rate and others. The most pronounced reactions in the ceramic material prior to the sintering process are related to the release of water. The release of water that is bound in the material occurs in different temperature regions as follows:

- *Physically bound water* (moisture) – water molecules are adsorbed on particle surfaces. The release from the material occurs in a temperature range of $(80 - 200)$ °C;

- *Crystalline water* – bound in stoichiometrically defined hydrates (e.g., $CaSO_4 \cdot 2H_2O$). The release occurs depending on the particular mineral (chemical compound) at temperatures of $(120 - 400)$ °C;

- *Structurally bound water* – is bound in the structure in the form of OH⁻ groups. It is found in most clay minerals such as illite, kaolinite, etc. It is released at temperatures of $(250 - 750)$ °C to form metaphase [22].

At higher temperatures, the breakdown of the structure of the original constituents and the formation of a new structure is characteristic.

1.4.2 The stages of firing

Depending on the reactions in the ceramic shard, the firing can be divided into several stages for which heating rates or isothermal soaking are to be considered. During rapid heating, temperature gradients are created in the body volume due to the slow temperature equalization between the surface and the interior of the body. Thermal expansion results in stresses which may, when exceeding a critical value, lead to the formation of micro-cracks or cracks, hence damage to the product [22].

Final drying. It is a continuation of drying, releasing physically bound water. It starts at 80 °C and, depending on the heating rate, the end shifts up to a temperature of ~300 °C. The heating rate should be chosen so that water vapor has enough time to diffuse to the body surface. Otherwise, increased pressure could lead to cracking and product damage.

Solid-phase reactions. This phase includes the process of heating (thermo-oxidation) of organic constituents (300 °C – 500 °C), dehydroxylation (450 °C – 700 °C), dissociation of carbonates (700 °C – 800 °C), and solid-phase sintering (800 °C – 900 °C). In these temperature ranges, volumetric changes occur due to thermal expansion and subsequent shrinkage of clay minerals and temper. Up to about 800 °C, when a more pronounced solid-phase sintering begins, the consolidation is low. The strength is influenced by the modification transition of the quartz and the release of CO_2, which can disrupt the grain contact areas or the individual grains.

Compaction. It occurs approximately in the temperature range from 800 °C to 1200 °C. A first melt appears, filling the pores. The pores and the body shrinks.

Sintering. There is considerable compaction of the shards. Shrinkage slows down. This area ends with reaching the firing temperature and isothermal soaking temperature. The proportion of melt increases and the body acquires thermoplastic properties. For this reason, the viscosity cannot drop below a certain threshold where the body would deform under its own weight. The sintering may take place solely in a solid state or in the presence of a melt. In silicates, sintering with the melt or a low viscosity glass phase are more common.

Cooling. It starts after reaching the maximal firing temperature and application of isothermal soaking. Until the temperature drops to ~800 °C, the body is not sensitive to the cooling rate, as the stresses are not generated due to a high plasticity in the presence of glass [22]. The reason for this is that the glass surrounding crystals has a relatively low viscosity if heated above a glass transition temperature. During further cooling, it is necessary to consider its rate, especially with regards to modification changes of SiO_2. In addition to the modification changes, the supersaturated melt crystallizes and the glass phase solidifies during cooling. The temperature range from ~500 to ~150 °C is also dangerous because micro-cracks can be formed due to different thermal expansions of the various phases of the ceramic shard [39].

1.4.3   Sintering

Sintering is the most important firing stage. During sintering, compacted bodies made of powders are densified by heat. It is one of the oldest human technologies originating in the prehistoric period [40]. During sintering, the original powder material gets a new arrangement, compacts (thus partially decreasing porosity), and strengthens. Simultaneously, its other properties, such as color and thermal conductivity, also change due to changes in the structure. Strengthening is achieved either by interconnecting the solid phase grains as they grow or by the melt-sealing effect. Along with sintering, other processes can also occur in the body to alter the phase composition, such as the formation spinel and mullite [21, 22].

In principle, sintering can be divided into two types: solid-phase sintering and liquid-phase sintering [40]. Solid-phase sintering occurs when the body is compacted exclusively in the solid phase at the sintering temperature, while liquid-phase sintering occurs when the liquid phase (melt) is also present during sintering. In addition to solid or liquid phase sintering, intermediate phase sintering and viscous flow sintering may also take place. Viscous flow sintering occurs when the melt volume is high enough to allow for complete compaction to be achieved by the viscous flow of the grain-melt mixture without altering the grain shape during densification. Sintering in the transition liquid phase is a combination of solid and liquid phase sintering. In this case, the melt is formed at the beginning of the sintering but disappears during compaction, and the sintering is completed in the solid phase. In general, liquid phase sintering makes it easier to control microstructure and to reduce manufacturing costs compared to solid-phase sintering. On the other hand, it may degrade important properties, such as mechanical strength [40].

The driving force of sintering is the reduction of the overall interface energy. The total interface energy is defined as γA, where γ is the specific surface energy (of the interface) and A is the total interface area. Total energy reduction can be expressed as [40]

$$\Delta(\gamma A) = \Delta\gamma A + \gamma \Delta A, \qquad\qquad (1.1)$$

where the change in interface energy ($\Delta\gamma$) is due to densification. The change in the interface area ΔA is the result of grain growth. For solid-phase sintering, $\Delta\gamma$ is given by the substitution of the solid phase-vapor interface for the solid phase-solid phase interface [40].

2.   Illite

Clays are the primary materials used in the production of traditional ceramics. The main components of such clays are mostly minerals of kaolinite and illite. The knowledge of high-temperature transformations of these minerals is fundamental to the understanding of the structural and microstructural properties of the final ceramic product [41]. Here we shall consider clay mined near the city of Füzérradvány (Tokaj region, Hungary). Several papers studying illite from this location were published in the past [18, 41, 42].

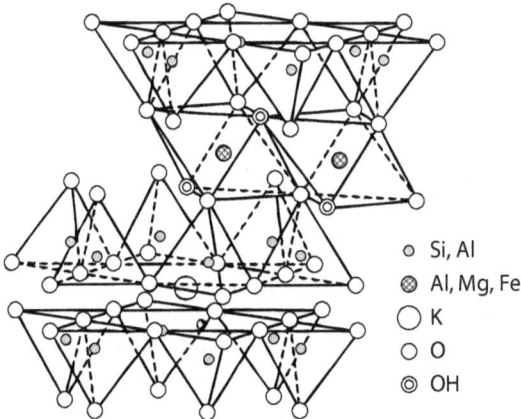

Fig. 2.1 – An ideal structure of illite (modified from [43]).

Frequent occurrence of illite is due to its high stability under standard conditions [19]. The term "illite" was introduced in 1937 by Grim et al. [44] as a general name for clay-sized minerals of the mica group often found in argillaceous sediments. The name was derived from the name of the state of Illinois, where it was first located. Środoń and Eberl [45] used a more precise definition in 1984, referring the term "illite" to a nonexpanding, dioctahedral, aluminous, potassium mica-like mineral. Illite, together with kaolinite, smectite, and chlorite, is one of the major components of sedimentary rock [41, 45]. It belongs to the aluminosilicates with a layered structure in the group of minerals of illite (together with hydromuscovite), which is a subgroup of clay minerals. Illite crystallizes in a monoclinic system in the C2/c spatial group [46], and its structure is very similar to mica and muscovite. Its structure consists of repeated silica tetrahedron – alumina octahedron – silica tetrahedron (TOT) layers [19]. The ideal structure of illite is shown in Fig. 2.1. According to the protocol of the International Mineralogical Association on nomenclature for the micas, potassium mica containing interlayer potassium cations in the range of 0.85 to 1 per half of the unit cell is defined as muscovite, while potassium mica having an interlayer cation in the range of 0.6 to 0.85 in half of the unit cell is defined as illite [47].

In addition, illite contains more silicon, magnesium iron, and water. The chemical formula of illite is [19]

$$(K, H_3O)(Al, Mg, Fe)_3(Si, Al)_4O_{10}[(OH)_2, (H_2O)] \,, \tag{2.1}$$

or in oxides form [48]

$$nK_2O \cdot Al_2O_3 \cdot 3SiO_2 \cdot nH_2O \,. \tag{2.2}$$

In [49] an approximate chemical formula for illite is given as

$$K_{0.88}Al_2(Si_{3.12}Al_{0.88})O_{10}(OH)_2 \,. \tag{2.3}$$

The tetrahedral layer consists of a network of $[SiO]^{4-}$ tetrahedrons connected to a hexagonal 2D structure. This layer is also referred to as a *silicate* layer. Approximately 20% of Si atoms in the tetrahedral layer are replaced by Al atoms. The octahedral layer is similarly formed by a 2D hexagonal network of octahedral cells. This layer is also referred to as a *gibbsite* layer.

In nature, illite appears in several polytypes[1]: the same polytypes observed in muscovite (1M, 2M$_1$, 2M$_2$, 3T) and a completely disordered turbostratic polytype 1M$_d$ [49, 50]. Illite often occurs in the form of mixed layers, especially with smectite [51, 52]. The 1M and 2M$_1$ polytypes are the most frequently used illite polytypes in the production of traditional ceramics [41]. Illite is a mineral often applied in mixtures for brick production due to its better plasticity as compared to kaolinite [48]. The SEM in Fig. 2.2 shows the flake-like illite crystals.

Fig. 2.2 – SEM of illite mined in Hungary.

---

[1]Polytypism is a special case of polymorphism where two modifications differ only in the way of stacking the identical 2D layers. The 1M polytype indicates that the individual layers are not twisted relative to each other, 2M polytype would indicate that every second layer is identical. More information about polytypism can be found e.g. i [110].

2.1   Origin of illite

Illite is formed by igneous or metamorphic rocks and their derivates. It is created in a high pH environment and, due to its sandwich 2:1 TOT structure, it requires an environment with an abundance of silicates and aluminates. The presence of an increased amount of potassium is also essential. Weathering of feldspar in cold conditions often leads to the formation of illite. An essential source of illite is also the alteration of muscovite, montmorillonite, or smectite [52]. In [53] the authors describe the illitization process (formation of illite from smectite) as a growth controlled by surface diffusion. The illite particles with a thickness of 2 nm are formed in the interlayer of smectite.

The structural stability of illite minerals accounts for their resistance to weathering and thus relative abundance in different soils and sediments [19]. It occurs as aggregates of small monoclinic grey to white crystals. Due to a small size of illite crystals, positive identification usually requires XRD or SEM-EDS analysis [54].

2.2   Development of illite structure during thermal treatment

During thermal treatment, the structural changes of illite occur that directly affect the properties of a body made of a material containing a high amount of the illite. That is the reason why the study of high-temperature processes in illite, as one of the main compounds used in the production of traditional ceramics, is essential.

In the study [38], high-temperature phase changes of illite were studied using *in-situ* XRD analysis up to the temperature of 1400 °C. It was concluded that the illite lattice undergoes dehydroxylation in the temperature region from 475 °C to 600 °C. The resulting anhydrides begin to disappear at 850 °C. The major high-temperature phases which are formed from illites are β-quartz (1000 °C – 1300 °C), spinel (1000 °C – 1400 °C), and mullite (1150 °C – 1400 °C). At temperatures above 1000 °C, a minor amount of cordierite, calcilite, feldspar, corundum, iron oxide, and forsterite may also occur. Bohor in [38] postulated a hypothesis that the degree of electronegativity and ionic potential of the adsorbed cations is proportional to the rapidity and degree of mullite crystallization. The reason for this is the influence of these factors in ion transfer and reaction product removal in the critical liquid phase at the clay mineral surface. Certain adsorbed cations act as repressors of high-temperature phase development. The order of decreasing repressive effect is [38]

$$K^+ > Na^+ > Li^+ > Ca^{+2} > Mg^{+3} > Be^{+2} > Sn^{+2,4}. \qquad (2.4)$$

Furlong [55] studied high-temperature processes of illite using *in-situ* transmissive electron microscopy. In illite samples, no observable changes took place until the temperature reached about 700 °C. At this temperature, a liquid phase started to form. The observable changes began within the temperature range of the anhydrite structure. The microdiffraction up to this point did not reveal any changes in patterns representing the *a-b* plane of illite. At 1000 °C, no reflections could be obtained, and no further diffraction maxima were obtained until the appearance of spinel patterns at temperature ~1100 °C. After 40 minutes at temperature 1150 °C, decomposition in the solid-state in some areas occurred. In these areas, intense spinel reflections were recorded with weaker mullite reflection. In other areas, partial or complete melting occurred, and diffraction patterns showed intense development of mullite with occasional spinel reflections present. At temperature 1200 °C, the active development of mullite crystals was observed. During dehydroxylation, the OH groups are lost, which leads to the disruption of two tetrahedral silica layers. Some of the disrupted tetrahedra have to be removed for to spinel phase to be formed. In this way, Furlong explains the creations of areas with the liquid phase. He correlated the hypothesis of liquid (melt) expulsion and the loss of the anhydride structure to the second endotherm, observable on DTA record, in the temperature range (~900 °C – 1000 °C).

The kinetics of illite dehydroxylation was studied by Gualtieri and Ferrari in 2006 [41]. The illite which was used in their work was from the Füzérradvány (Hungary). Results of a chemical analysis of the studied illite sample are in Table 2.1. The authors pointed to the importance of a comparison of their results with results obtained for muscovite due to their very similar structure. To separate illite from the other admixtures in raw illitic clay, the sedimentation method was used. Phase changes during thermal treatment were recorded by a diffractometer. The conversion coefficient was also calculated using the area of selected peaks. The time dependence of the conversion coefficient was the basis for the kinetic analysis by the isoconversional and Avrami method. The direct application of kinetic equations was used to verify the mechanism of the reaction. Referring to their previous work, Gualtieri and Ferrari assumed that the investigated illite sample was the 1M-polytype – likely a disordered $1M_d$-polytype.

Table 2.1 – Chemical composition of illite in wt.% (L.O.I. – Loss on ignition) [41].

SiO_2	Al_2O_3	Fe_2O_3	CaO	MgO	TiO_2	K_2O	Na_2O	P_2O_5	L.O.I.
50.2	0.39	0.33	0.24	3.451	0.08	7.03	0.17	0.07	8.04

According to the results of isoconversional method, the authors supposed that dehydroxylation of illite is a two-step reaction. This result is supported by DTA analysis, where, in the temperature interval from 600 °C to 700 °C, two endothermic peaks were observed. Also, the result of the Avrami method confirmed at least a two-step dehydroxylation process. An Arrhenius plot was used for the calculation of activation energies of 697 kJ mol^{-1} and 231 kJ mol^{-1} for the first and second steps, respectively. Using the model-independent method, the activation energy values of the two-step process were

676 kJ mol^{-1} and 230 kJ mol^{-1}, respectively. According to the results, Gualtieri and Ferrari report that the dehydroxylation model is consistent with those present in the literature for muscovite and other 2:1 phyllosilicates. Therefore, based on a previous work, for illite they proposed a multi-step model of dehydroxylation (the first stage is divided into separate two steps) composed of:

- Condensation of water molecules, from OH groups, in the octahedral layer according to the reaction $2(OH) \rightarrow H_2O + O$,

- 1D diffusion of water molecules throughout the tetrahedral ring enlarged by thermal expansion,

- 2D diffusion of water molecules through the interlayer region.

The reaction rate is limited by the 1D and 2D diffusion processes of water molecules. The first step (condensation of water molecules) could not be distinguished in their diffraction measurements. Thus, they claim that the activation energy of the first step (676 kJ mol^{-1}) involves the condensation of water molecules as well as 1D diffusion throughout the tetrahedral ring. According to the authors, most of the energy is consumed by 1D diffusion. 2D diffusion of water molecules through the interlayer region was possible due to the vacancies of K$^+$ ions. In previous studies, two-step dehydroxylation was not recognized; therefore, the apparent activation energy values were reported in the range of 325 kJ mol^{-1} – 400 kJ mol^{-1}.

Phase changes on a Hungarian illite 2:1 clay were studied by Carroll et al. in 2005 using ^{1}H, ^{27}Al, ^{29}Si, ^{39}K solid-state nuclear magnetic resonance technique (NMR) [42]. Illite was studied on samples heated up to a temperature of 1600 °C. The studied illite was single-phase with a small amount of aluminum substitution in the silica layer and low content of Fe$_2$O$_3$ (0.4 wt.%). Compared to other works, a low content of iron allowed for obtaining better results by NMR. The chemical composition of studied Hungarian illitic clay is in Table 2.2.

Table 2.2 – Chemical composition of Hungarian illitic 2:1 clay in wt.%
(L.O.I. – Loss of ignition) [42].

SiO$_2$	Al$_2$O$_3$	Fe$_2$O$_3$	CaO	K$_2$O	LOI
49.6	33.0	0.4	1.6	6.6	8.1

The authors observed an increasing number of AlO$_4$ and AlO$_5$ bonds during dehydroxylation. They also reported that silicate and gibbsite layers were separated during dehydroxylation. In the temperature range from 900 °C to 1000 °C, silicate layers formed potassium-aluminum-silicate glass. Gibbsite layers formed a spinel at this temperature and also a small amount of aluminum-rich mullite. At 1600 °C, the authors observed an increase in the aluminum content of the potassium-aluminum-silicate glass. They also reported an increased mullite fraction and formation of $\alpha - Al_2O_3$.

2.3   The use of illite in the production of building ceramics

Although illite is the main component of clay soils used for traditional ceramics production, the influence of illite on the properties of ceramic bodies has been investigated only in a relatively small number of studies.

Ferrari and Gualtieri [18] focused on quantitative mineralogical aspects of unfired and fired ceramic bodies. They also studied their relation to technological properties influenced by the presence of illite. Raw materials with different amounts of illite were used in their study. Illite originated from different locations, one of them being Füzérradvány. The chemical composition of this material was very similar to that in Table 2.1. The authors found out that a higher amount of illite increased the glassy phase portion, decreased the temperature of the glass transition and, possibly, absorption of water. The presence of illite also initiated the cristobalite and mullite formation. The color of bodies fired from this material was darker despite a low content of transition metals in the Hungarian clays.

The influence of the amount of illite on the phase composition of ceramic bodies after firing was also studied by Aras [56]. The potassium content in illite-containing mixtures increased the glassy phase and mullite formation. The formation of cristobalite is completely suppressed in the presence of illite, which is slightly contradictory to the results of Ferrari and Gualtieri [18].

High-temperature processes during the sintering of illite were investigated by Wattanasiriwech et al. [57]. The authors concluded that vitrification of illitic clay started at around 900 °C, and full densification was achieved at 1200 °C without the addition of feldspar as a commonly used melting agent. In this case, illite also acted as a melting agent and contributed to eliminating the pores. Results of the XRD analysis showed that the illite transformed to sericite below 900 °C, and the sericite reflections completely disappeared at 1000 °C. Melting started at 1000 °C. The SEM showed long mullite crystals grown from the edges of the liquid regions toward the inside of the molten region. The spinel formation was not observed at all. This result was inconsistent with the conclusions of Caroll et al. [42]. Wattanasiriwech et al. [57] concluded that illite suppressed the formation of spinel and shifted mullite formation to lower temperatures.

In [58] several clays were analyzed. One of them, according to the XRD analysis, contained 72 wt.% of illite, 15 wt.% of kaolinite with feldspar, and quartz as minor phases. The content of particles with a size below 2 µm was 20 wt.%. The samples were prepared by pressing at a pressure of 3 MPa. The determination of firing characteristics in the temperature range of 900 °C – 1250 °C was performed at 200 °C/h, where the samples were maintained at the maximal temperature for 2 hours. The results of the measured quantities after firing at different temperatures are shown in Table 2.3.

Table 2.3 – The properties of bodies with 72 wt.% of illite fired at different temperatures [58]. RE – relative expansion (in %), WA – water absorption (in %), and BD – bulk density (in g cm⁻³).

Quantity	Firing temperature / °C							
	900	950	1 000	1 050	1 100	1 150	1 200	1 250
RE	0.42	0.12	−1.41	−4.75	−7.55	−8.57	−5.90	−4.24
WA	17.70	17.09	13.88	7.79	1.51	0.00	0.02	0.02
BD	1.78	1.75	1.84	2.13	2.40	2.48	2.08	1.95

3.   Power plant fly ash

Power plant ash is waste from the combustion of solid fuels in thermal power plants [3]. This fuel is mostly coal, but modern thermal power plants can also burn municipal waste, biomass, and other combustible waste from industrial production [59–61]. In this work, attention will be paid to the power plant fly ash resulting from the combustion of brown coal or a combination of brown coal and a minor part of wood chips.

Coal combustion in thermal power plants is currently based on two leading technologies – pulverized coal combustion (PCC) and fluidized bed combustion (FBC) [3]. For the needs of PCC boilers, coal is first ground to dust (particles up to ~0.2 mm) using fan mills. The powder is then injected into the combustion chamber.

PCC boilers use a combustion temperature in the range of 1100 – 1300 °C in order to avoid the formation of large quantities of liquid slag. Despite this, some parts of the softened ash particles bond together creating a slag that falls to the bottom of the furnace. From there, it is removed mechanically or hydraulically. Lignite (brown coal) or black coal is used.

Combustion in FBC boilers [3, 62] is based on the effect of fluidization (liquefaction) of a mixture of coal, ash, sand, limestone, etc. Such a mixture acquires the properties of a "boiling" liquid by injecting air into it from below. Its temperature, and thus its combustion temperature, is around 800 °C. Coal is fed into this mixture with a grain size of ~6 – 10 mm, and its preparation for combustion is, therefore, less demanding than in the case of PCC. Limestone or dolomite is added to the mixture to bind the sulfur oxides. The advantage of a low combustion temperature is that only a minimum of nitrogen oxides are formed. These advantages of fluidized bed combustion eliminate the need to build additional equipment to remove nitrogen and sulfur oxides from the flue gas in order to meet legislative limits. In addition, this technology makes burning of low-quality coal, oil shale, biomass, municipal waste, and others easier. Due to the advantages mentioned above and the fact that the combustion efficiency is higher than with other technologies, fuel combustion technology in fluidized bed furnaces is booming.

Coal ash is usually stored near a thermal power plant in heaps, repositories, or sludge ponds [3]. These landfills occupy relatively large areas (hundreds of hectares), which will lose agricultural use. Such fly ash storage brings considerable problems associated with groundwater and environmental contamination risk. Furthermore, fly ash contains toxic substances, especially heavy metals, e.g., Hg, Pb, As, Cd, Mo, Cr, Sr, and others [63]. Last but not least, the dustiness that arises when the ash is stored in the open, non-consolidated repositories should also be mentioned. Slightly increased radioactivity due to the presence of radionuclides can be an obstacle to the reuse of fly ash [4]. In addition, all the costs associated with storing fly ash are reflected in the price of electricity. This is one of the reasons why finding other options for power plant fly ash usage is significant.

The terminology used to describe coal combustion products varies among different resources. However, some definitions suggested by the World Wide Coal Combustion Products Network is as follows [64]:

Coal combustion products – include fly ash, boiler slag, fluidized-bed combustion (FBC) ash, flue gas desulfurization material.

Fly ash – finer ash produced in coal-firing power stations. It creates the majority (80 – 85 %) of non-combustible materials. It is transported by combustion gases (flue gas) and captured at the boiler outlet (usually by mechanical and/or electrostatic precipitators).

Bottom ash – coarse ash that falls to the bottom of a furnace, also known as furnace bottom ash. About 15 % of the ash production.

Cenospheres – hollow ash particles that form in the furnaces gas stream. About 1 – 2 % of the ash production.

Conditioned ash – fly ash is mixed with water (10 – 20 % by dry mass) to be transported in normal tipping vehicles.

Flue Gas De-Sulfurisation – a source of Calcium is injected into the furnace gas stream to remove sulfur compounds.

3.1 Coal

The ash properties are determined by the type and composition of the coal from which it was formed, its combustion technology, and its temperature history [3]. Therefore, basic information on coal will be provided briefly at this point.

Coal is a black or brown-black combustible rock formed during the charring of dead plants by the action of heat and pressure. According to the degree of charring, we divide coal into [3, 65]:

- **Anthracite** – is a hard, brittle, and black lustrous coal, often referred to as hard coal, containing a high percentage of fixed carbon (86 – 98 %) and a low percentage of volatile matter.

- **Bituminous** – is middle-rank coal between subbituminous and anthracite. It is blocky and appears shiny and smooth, but thin, alternating, shiny, and dull layers can be seen on a closer look. Bituminous contains 69 – 86 % of carbon by weight.

- **Subbituminous** – is black and is mainly dull (not shiny).

- **Lignite** – also known as brown coal, has a low heating value and high moisture content (up to 70% water by weight) and is mainly used in electricity generation.

The elements carbon, oxygen, hydrogen, sulfur, and nitrogen are the most abundant in coal. Coal consists of three components: physically bound water, volatile matter, and nonvolatile matter with a proportion of combustible components and fly ash. In fly ash, in addition to the volatile component and moisture, the substances contained in the coal are retained but at a higher concentration (2 to 3 times higher after burning lignite). In Slovakia, subbituminous coal and lignite are mined. Almost 50 % of imported coal is bituminous [3].

The estimated amount of the world's coal reserves are 850 Gt, which would last for another 130 years [64]. China, USA, India, Australia, and Russia are the largest coal producers [64]. The worldwide production of seam and coking coal in 2020 was 6.97 Gt in addition to 0.64 Gt of lignite [66]. The bulk of the mined coal is consumed in the country of origin for energy production, primarily by pulverized coal combustion technology [64]. Globally, coal is the essential fuel used for energy production. Currently, around 30% of primary energy (heat) and 40% of electricity is generated globally by coal combustion [67]. Therefore, the demand for coal in the near future is expected to rise in countries like India, Southeast Asia, and a few other countries in Asia. On the other hand, coal demand is expected to decline in Europe, Canada, USA, and China. Therefore, the global coal demand will probably only increase slightly in the next decade [67]. The decrease in coal demand in developed countries is driven by demands for climate change mitigation, transition to renewable energy forms, and increased competition from other resources such as natural gas and oil [67].

3.2 Formation of power plant fly ash

The combustion of coal in thermal power plants produces large quantities of solid products, often referred to as waste. However, they should more accurately be defined as coal combustion products (CCPs), a term that is more positive and adheres to the concept of industrial ecology, i.e., an approach that transforms a by-product of one industry into raw material for another industry [64].

To fully understand the properties of fly ash, it is also necessary to know the conditions of its formation. Depending on the type of coal used, fly ash makes up about 25% of the weight of the original coal. Of this, about 65% are very fine particles with a diameter of up to 100 μm [68].

Fly ash is produced in three phases in PCC boilers [3, 69]:

    1. melting of the mineral components in the coal,

    2. agglomeration of molten ash droplets,

    3. fly ash formation.

The formation and character of ash are influenced by the kinetics of combustion, which depends, in addition to chemical processes, on a set of physical conditions determined by air supply, diffusion phenomena, heat supply, size and surface properties of coal particles, its thermal conductivity, etc. [3].

The ashes are usually polycomponent, i.e., the individual particles have different compositions (chemical and mineralogical). Fly ash from FBC boilers (FFA) retain the morphology of the original coal grain, are porous, and contain a low proportion of glass phase, depending on the content of low-melting eutectics in the ash [3]. FFA has a high chemical activity due to its large specific surface area. In contrast, fly ashes from other boiler types contain a significant proportion of glass phase, and their surfaces are more smoothed. Not all coal burns thoroughly in the combustion process. Therefore, its content can play a crucial role in further using the fly ash [3, 70].

The flue gases transport the fine ash produced during coal combustion to the backdrafts of the chimneys, where it is mechanically and/or electrostatically separated to reduce its emissions to the environment. The fly ash is then transported for further processing or to storage sites.

3.3   Power plant fly ash types

According to Heidrich et al. [64], either calcareous or siliceous fly ashes are generated depending on the coal used. In siliceous fly ashes, three elements are dominant: silicon, aluminum, and iron. The oxide content is in the range of 75 % – 85 %. They consist mainly of glassy spheres together with some crystalline phases and unburnt carbon. By definition, the amount of CaO is limited to less than 10 %. Calcareous fly ash contains the same oxides, but the CaO content is higher than 10 %. Different countries use different definitions, for example, the European standard, EN 450-1 [71], the Australian standard AS3582.1 [72], or others. Essential parameters of fly ash that are specified in the relevant standards are carbon content, chemical, and mineralogical composition, etc. According to ASTM C 0618-03 [73], fly ash can be classified according to composition into three categories (Table 3.1).

Table 3.1 – Classification of power plant fly ash according to standard ASTM C 618-03.

	Class		
	N	F	C
Minimum proportion of $SiO_2 + Al_2O_3 + Fe_2O_3$ / %	70	70	50
Maximum amount of SO_3 / %	4	5	5
Maximum moisture content / %	3	3	3
Maximum ignition loss / %	10	6	6

3.4   Chemical and mineralogical composition of fly ash

The properties of fly ash are primarily conditioned by the properties of the non-combustible components (fly ash) of the original coal grain. Silicates, carbonates, and sulfides predominate among the minerals [3]. Chemical changes during coal combustion have the character of oxidation and reduction reactions. Another important factor influencing the mineralogical composition is the combustion temperature. Carbonates dissociate even at temperatures above 700 °C, forming CaO or MgO oxides. They bind sulfur to form the corresponding sulfates [3]. At temperatures above 1200 °C, mullite, cristobalite, or quartz are formed from the original clay minerals. Combustion temperatures above 1300 °C can lead to the formation of a silica melt, which passes into an amorphous phase by rapid cooling in chimneys. The crystalline phase content in the ash from the PCC boilers is less than 15% [3]. Table 3.2 shows the typical chemical composition of fly ash from PCC boilers depending on the coal used (taken from [64]).

Coal combustion technology in fluidized bed boilers requires the addition of calcite (or dolomite). The calcium oxide produced by calcination is designed to bind sulfur to itself and thus reduce the sulfur content in the flue gases leaving the chimney. It is added at about twice the stoichiometric ratio to desulphurize the products after burnt coal. This is reflected in the increase in the content of free CaO or $CaCO_3$ in the ash from fluidized bed boilers.

Table. 3.2 – Typical chemical composition of fly ash from pulverized coal combustion boilers (PCC) according to [64] (in wt.%). L.O.I. – Loss on Ignition.

Oxide	Bituminous coal	Subbituminous coal	Lignite
SiO_2	20-60	40-60	15-45
Al_2O_3	5-35	20-30	10-25
Fe_2O_3	10-40	4-10	4-15
CaO	1-12	5-30	15-40
MgO	0-5	1-6	3-10
SO_3	0-4	0-2	0-15
Na_2O	0-4	0-2	0-6
K_2O	0-3	0-4	0-4
L.O.I.	0-15	0-3	0-5

3.5   Ash from Nováky power plant

The ENO Nováky thermal power plant burns lignite mined in Slovakia in the mines of Nováky, Čáry, Cígeľ and Handlová. The calorific value of this coal is in the range of (9.88 – 10.30) MJ kg^{-1} [2].

Materials Research Foundations **130** (2022)   https://doi.org/10.21741/9781644902073

The Nováky power plant is divided into ENO A and ENO B sections. In the ENO A section, one FBC boiler is in operation. It burns on average 800 t of coal (<20 mm) per day together with (80 – 150) t of limestone (<10 mm). In addition, (50 – 60) t of wood chips are burnt daily with coal. The addition of limestone achieves desulphurization of the coal ash, with the sulfur then being bound in the ash. In the ENO B section, four units were in operation by the end of 2015. Currently, only ENO B units 1 and 2 are in operation, each with one PCC boiler and an installed turbine capacity of 110 MW. The dried coal is ground in fan mills to a maximum particle size of 0.2 mm before firing. The ash produced by combustion is first ionized and then collected by electrostatic precipitators, where a separation efficiency of up to 99.9 % is achieved.

Nováky power plant produces three types of fly ash:

- *Fluidized bed combustion Bottom Ash (FBA)* - ash from the fluidized bed boiler (ENO A Nováky) of coarse granularity containing particles (mainly quartz) that were too heavy to be transported with the flue gases to the rear flue drafts. It also contains coarse impurities.

- *Fluidized combustion Fly Ash (FFA)* - fly ash from the fluidized bed boiler (ENO A Nováky) of fine fraction, transported from the heating site by flue gases to the backdrafts of the chimney, where electrostatic precipitators separated it. Part of it was returned to the heating plant, where it co-formed the fluidized bed.

- *Pulverized firing Fly Ash (PFA)* - fly ash of fine fraction from pulverized coal combustion boilers (ENO B Nováky, blocks 1 and 2). It was separated from the flue gas by mechanical and electrostatic precipitators.

After the ash is separated by electrostatic precipitators or collected from under the grates, the ash from the ENO B section is hydraulically transported to the ash ponds, the total area of which is ~163 ha. At ENO Nováky, the fluidized fly ash is almost entirely used for the production of stabilizers for existing fly ash landfills. Nevertheless, in this work, we will investigate the possibilities of using this fly ash to produce ceramic materials. The current trend is the introduction of fluidized bed combustion, and hence an increased production of this type of fly ash can be expected in the future. The rate of reuse of fly ash from the PCC boiler (PFA) in ENO Nováky in 2014 was 52%, with a total production of 445.4 t and external removal of 232.7 t. Of this, the most significant part was used to produce aerated concrete, the rest as an additive for concrete and cement. In addition to the three types of fly ash mentioned above, ENO B Nováky also produces bottom ash (slag) during the pulverized coal combustion process, which is hydraulically transported to the ash pond. The content of fly ash in slag is about 20 %. Part of the slag is taken by brick manufacturers (26 t in 2014). The fly ash used in this research was collected directly after separation in the separators, prior to trans-shipment to the repository.

3.6 Utilization of power plant fly ash

The use of fly ash began with the advent of pulverized coal combustion technology in the 1920s [64]. In 2016, 1222 Mt of coal combustion products were generated, of which 64% found secondary use [67]. Comparing to data from 2010, both global production and utilization rate increased from 780 Mt and 53%, respectively [64]. CCPs utilization rate varies considerably from country to country (Table 3.3). Japan is the most efficient user of CCPs, followed by Europe (EU15) and Israel.

Table 3.3 Estimated annual production and utilization rates of CCPs in 2016 according to [64].

Country/Region	CCPs volume / Mt	CCPs utilization / Mt	CCPs utilization / %
Australia	12.3	5.4	43.5
Asia			
–China	565	396	70.1
–Korea	10.3	8.8	85.4
–India	197	132	67.1
–Japan	12.3	12.3	99.3
–Other Asia	18.2	12.3	67.6
Europe	140		
–Europe (EU15)	40.3	38	94.3
Middle East and Africa	32.2	3.4	10.6
Israel	1.1	1	90.9
USA	107.4	60.1	56.0
Canada	4.8	2.6	54.2
Russian Federation	21.3	5.8	27.2
Total	**1 221.9**	**677.7**	**63.9**

Based on the European Coal Combustion Products Association data, 40 333 kt of CCPs were generated in the EU15 countries in 2016. The utilization rate, excluding fly ash for reclamation and filling old opencast mines and pits, was 50.1%. With the inclusion of CCPs for these purposes, the utilization rate was as high as 94.3%. Data on the specific use of CCPs from the same source are given in Table 3.4. After reclamation, the most significant part is used as an additive in concrete. Here, fly ash's hydration and pozzolanic properties are exploited [4]. The definitions and related properties of CCPs for such purposes are covered in national standards, e.g., EN 450-1 and ASTM C 618. A large part of the CCPs is used to produce plasterboards; it is not directly fly ash but flue gas desulphurization products. The utilization for ceramics production was only 0.21%, and according to more detailed data, 74% of this is fly ash, and 26% is bottom fly ash.

Table 3.4 CCPs utilization in Europe (EU15) in 2016.

	Utilization	
	kt	%
Mine reclamation	17 239	42.7
Concrete additive	4 652	11.5
Plasterboard	4 173	10.3
Cement production	2 076	5.1
Filling	2 502	6.2
Leveling screed	1 356	3.4
Ceramics	85	0.21
Other	5 945	14.7
Total	**38 028**	**94.3**

Other possibilities of utilization of fly ash can be found in the literature; such as the use of fly ash in agriculture, a fertilizer or additive to reduce soil acidity [1], for the production of glass and glass-ceramics [5, 74, 75], geopolymers [6, 76, 77], zeolites [4, 78, 79] mullite, composite materials with plastics and metals, and heavy metal absorbers [78]. Also of interest is the possibility of separating functional components, mainly metals Fe, Al, and Ti [3, 4, 78] but also Ga, V, Ni, Mg, and others [78].

Utilization of power plant fly ash for the production of building ceramics

The ash has a chemical and granulometric composition close to the raw materials used to produce traditional ceramics (clay minerals). Therefore, it is assumed that it could be used as a suitable substitute [14]. Efforts to produce construction ceramics from fly ash can be divided into two groups based on a study of the literature:

- sintering of the fly ash
- replacing some of the raw materials of traditional ceramics with fly ash

The possibility of using fly ash as an additive in building ceramics has been a research topic for about 20 years. The following section will overview the works that deal with this issue.

In [80], Acar and Atalay discussed the possibility of producing ceramics from 100% fly ash from pulverized coal combustion (without any other additives). Samples produced by dry pressing were fired at temperatures ranging from 1000 °C to 1150 °C with dwell times between 0.5 and 2.0 hours. Optimum properties were obtained when the fly ash was sintered at 1150 °C with a 1.5 h dwell time. Sintering decreased the proportion of the amorphous phase.

A similar approach was followed by Erol et al. in [81]. They produced samples from fly ash by pressing without any additives, which they fired at temperatures ranging from 1025 to 1175 °C. The properties of the obtained materials were dependent on the sintering temperature, dwell time, the grain size distribution of the fly ash, the mineralogical and chemical composition of the fly ash, and pressing pressure. The higher the $Al_2O_3 + SiO_2$ content was, the better the properties of the sintered materials were. The densities of the sintered materials are comparable to commercially produced ceramics. SEM analysis showed the formation of needle-like mullite crystals in the material structure. The results of the XRD analysis showed that the main mineralogical phase after firing was mullite. The presence of anorthite was also observed, which correlated with the CaO content of the original fly ash.

The same authors compared the sintered fly ash with the glass and glass-ceramics made of the same fly ash [75]. The dominant mineralogical phases are enstatite and mullite in both sintered fly ash and glass-ceramics. The microstructure for glass, glass-ceramics, and ceramics made from the same fly ash differs due to the different preparation methods. Glass and glass-ceramics had better physical properties than sintered fly ash (ceramics). They reported that the heavy metals contained in fly ash were successfully bound in the material by sintering.

In [82], a partial description of the preparation of bricks from pure fly ash and unspecified additives by a patented technology is given. The firing temperature was in the range of (1000 – 1300) °C with a dwell time of several hours. 2.75 tons of fly ash are needed to produce 1 000 pieces of bricks. The bricks have a lower bulk density and a higher mechanical compressive strength than standard bricks (43 MPa compared to the typical value for bricks of (12 – 40) MPa).

Olgun et al. [83] found that replacing potassium feldspar (whose content was 10 wt.%) with fly ash in a mixture for the production of tiles increased their flexural strength by 68% and increased the shrinkage by firing from 2.53% to 3.85%. The addition of fly ash further causes less homogeneity in pore distribution and overall microstructure.

In [14], Zimmer and colleagues focused on using fly ash from coal combustion to produce tiles. The samples were produced by uniaxial pressing at a pressure of 20 MPa. They experimented with mixtures with fly ash contents up to 80 wt.%. The plastic component consisted of kaolin clay. The optimum fly ash content was determined to 60 wt.% and the ideal firing temperature of 1150 °C.

The problem of secondary use of waste after brown coal combustion in the Czech Republic was addressed by Sokolář et al. [12]. They mainly focused on the influence of the grain size distribution of fly ash from PCC boilers. They used a mixture of 70 wt.% fly ash and 30 wt.% clay (with the main mineralogical phases being illite, feldspar, mica, and quartz). They monitored the effect of firing temperature in the range 1000 °C – 1150 °C. The samples were produced by uniaxial pressing at a pressure of 20 MPa. Unground fly ash allowed to obtain the same frost resistance at higher absorption as in the case of using ground fly ash. Grinding the fly ash reduced the firing temperature by approximately 150 °C. The use of ground fly ash and firing the mixture at 1150 °C increased the mechanical strength (34 MPa compared with 24.9 MPa for unground fly ash) but, on the other hand, also increased the unwanted shrinkage by firing (14.3% compared with 7.2%).

The effect of FFA substitution in fly ash-clay mixtures on the flexural strength was investigated in [13]. Conventional fly ash from PCC boilers (PFA) was substituted with FFA. The mixture always contained 60 wt.% fly ash and 40 wt.% clay (~65% kaolinite, ~25% illite). Samples were produced by uniaxial pressing of the moistened mixture (12% water) and with 0.2% pentasodium triphosphate as deflocculant. The upper limit of FFA in the mixture was determined to 15% at a maximum temperature of 1080 °C and a dwell time at this temperature of 10 min. The addition of FFA caused a decrease in the flexural strength of the green bodies from 2.8 MPa to 1.7 MPa, reduced the shrinkage by firing, decreased the flexural strength, and increased the water absorption. Overall, FFA deteriorates the parameters of the ceramic bodies. Partial improvement occurs if the FFA is ground, but this is an economically disadvantageous process. The work also points to the danger of releasing sulfur dioxide, which was significantly more abundant in the effluent from the firing of the FFA-containing samples.

The results of the study [1] suggest that fly ash ceramic mixtures are suitable for tile, tile, or brick production. The ceramic mixtures were fired at temperatures ranging from 900 °C to 1200 °C. The authors reported up to 50% of mullite after firing. The beneficial aspects of replacing traditional raw materials with fly ash are the reduction in milling costs, the reduction in sintering temperature, and the increase in frost resistance. Due to the different chemical compositions of fly ash from different sources, it is impossible to establish a general rule for its use in ceramics [1]. In [16], it was found that the addition of 5 wt.% of fly ash leads to a reduction in bulk density and an increase in the resistance of bricks to failure due to salt crystallization. Ceramic tiles containing 60 wt.% of ash fired at 950 °C achieved sufficient impact toughness and low shrinkage by firing [15]. In the study [10], up to 80% of the clays were replaced with fly ash when a compressive strength of 25 MPa was achieved. Moreover, the residual carbon used in the ceramic mixture can save part of the firing cost. The possibilities of using fly ash in pottery and artistic ceramics have been investigated in [8].

Michalíková et al. [2] discussed the possibility of using fly ash from the ENO Nováky power plant to produce ceramic tiles. FFA or PFA replaced a maximum of 12% of the original ceramic mass. PFA was evaluated as suitable for the purpose, but only in the case of slow firing. In the case of the use of fast-firing technology, the authors recommend the separation of iron oxide and coal fly ash components. Iron oxides cause a dark color of the

Materials Research Foundations **130** (2022)   https://doi.org/10.21741/9781644902073

ceramic shard and problems in the case of rapid firing technology (formation of reduction nuclei, expansion, and destruction). The main problem of fluidized fly ashes was the Ca^{2+} content, which causes problems in the flowing of the ceramic slurry and the high porosity of the ceramic bodies.

If we wanted to compare the results of these works, we would have to consider many factors, such as the proportion of fly ash used, variation in chemical composition, different mineralogical phases, different sample preparation methodology, firing regime, etc. However, most of the papers agree with the conclusion that fly ash is suitable for the production of building ceramics, namely for the production of tiles and bricks. Therefore, the aim should be to use as much fly ash as possible to make the process economically and environmentally advantageous. The safety analysis of fly ash products may also deserve attention, as they are known to contain significant quantities of toxic compounds. However, some studies have already indicated that in the sintering process, toxic substances, namely heavy metals, are safely incorporated into the structure of the ceramic body [59, 75].

Although the idea of replacing part of the traditional raw materials of the ceramic industry with fly ash is not new, the use of fly ash in the ceramic industry has still not reached its full potential. The reason is the hesitation on the part of producers and the lack of more extensive research, which would serve as a basis for the elaboration of case studies of individual ceramic factories near the sources of power plant fly ash. Furthermore, the chemical, mineralogical and granulometric composition of fly ash varies between producers, and thus fly ash from different sources needs to be considered individually.

4.　Experimental procedure

4.1　Preparation of raw materials

Before the actual production of the samples, the raw material must be processed to meet the requirements for granulometry, processing stability, reproducibility of the results, and others. Emphasis was placed on a uniform approach to the preparation of the material at different times during the four years of this research. This is important in terms of comparing the results obtained on samples prepared at different times.

4.1.1　Illite

As pointed out in Section 2.3, clays are used as a plastic component in building ceramics. These consist mainly of clay minerals such as illite, smectite, chlorite, kaolinite [18, 19]. They often also contain a significant proportion of feldspar and quartz [19]. Clays are a mixture of these minerals, and their composition varies from site to site. From the point of view of material research, it is advantageous to study simpler systems and to predict the properties of materials containing other components based on the results obtained [39, 84, 85].

In this work, a material with a major content of illite (hereinafter referred to as illite) was used as the plastic component of the experimental mixtures. It was mined near the town of Füzérradvány in the northeast of Hungary. It consisted of particles of the order of microns in size up to lumps with a diameter of about 50 mm. These had to be mechanically disrupted first to particles below 5 mm suitable for a ball mill. The material was repeatedly ground and sieved through a 100 μm sieve until the sieve residue was below 2 wt.% of the total mass. The material thus prepared was used to prepare experimental mixtures.

4.1.2　Grog

In the manufacture of ceramic products, a material is added to the mixture to reduce the plasticity of the ceramic mass (in the case of wet bodies) and to prevent violent volume changes during drying and firing [21, 22]. Due to the temperature gradients, such a sharp shrinkage does not distribute evenly throughout the body, resulting in stresses which, if exceeded the breaking strength of the material, may cause damage to the article. The part of the raw material mixture to prevent this undesirable effect is called *temper*. It utilizes the stability of this material in the temperature ranges used in the firing of ceramic bodies. Quartz sand, slag, chamotte, corundum, but also crushed parts of waste ceramic products (grog) are most commonly used for this purpose. The advantage of using crushed waste ceramic products is the compatibility of the composition with the rest of the material and hence the ability to form a monolithic compact structure. Temperature stabilization has occurred in this material, and there is no significant physico-chemical change at the next firing [22]. At the same time, compliance with the chemical composition of other

31

components is ensured. The disadvantage is the need for expensive crushing and grinding. In an attempt to simulate such an admixture, powdered illite with particles below 100 μm was fired to 1100 °C with a holding time of 90 min. In doing so, the illite partially sintered and again had to be mechanically disrupted in a mortar. In this case, such manual crushing was sufficient to have zero residue on a 100 μm sieve. The illite prepared in this way has undergone considerable changes in firing. It had a much lower absorbency and swelling compared to the crude illite.

4.1.3   *Fluidized combustion Fly Ash (FFA)*

FFA was first mixed with demineralized water in an attempt to remove free CaO. The water level reached several centimeters above the surface of the fly ash. The mixture of water and fly ash was left in the laboratory for two days to achieve a complete reaction between free calcium oxide and water. After two days, the excess water was poured out, and the wet ash was freely dried in the laboratory for one week. Thus, part of the calcium hydroxide $(Ca(OH)_2)$ was converted by reaction with air CO_2 into calcium carbonate according to the equation

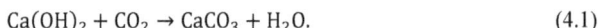

$$Ca(OH)_2 + CO_2 \rightarrow CaCO_3 + H_2O. \hspace{2cm} (4.1)$$

Part of the calcium carbonate could crystallize as calcite. Finally, the fly ash was dried at 120 °C and sieved through a 200 μm mesh screen. Only the fraction below 200 μm was used for sample preparation, which represented 65 wt.% of the original FFA fly ash (see screen analysis in Table 4.1).

Table 4.1 – Screen analysis of FFA after mixing with water and drying.

fraction / μm	<50	50-100	100-200	200-500	500-1000	>1000
amount / wt.%	0.9	14.8	49.3	26.3	6.1	2.6
fraction / μm	<50	<100	<200	<500	<1000	total
amount / wt.%	0.9	15.7	65	91.3	97.4	100

4.1.4   *Pulverized firing Fly Ash (PFA)*

This fly ash was also hydrated in the same manner as FFA prior to further use, although the predicted free calcium oxide content of CaO is substantially lower. The hydrated and dried PFA was sieved through a 200 μm sieve, and only a fraction below 200 μm was used to prepare the experimental mixtures. The overflow of the 200 μm sieve was up to 85.2 wt% (see Table 4.2). Thus, it can be said that the PFA fly ash is fine-grained and, when used in building ceramics, there is no need for energy-intensive grinding.

Table 4.2 – Screen analysis of PFA after mixing with water and drying.

fraction / μm	<50	<100	100-200	200-500	500-1000	>1000
amount / wt.%	-	55.5	29.7	13.9	0.54	0.36
fraction / μm	<50	<100	<200	<500	<1000	total
amount / wt.%	-	55.5	85.2	99.1	99.64	100

4.1.5   Fluidized bed combustion Bottom Ash (FBA)

This fly ash, like FFA, comes from block A of the Nováky power plant where it uses the technology of burning brown coal (and about 6% wood chips) in a fluidized bed. Since it is bottom ash and not fly ash, it contains coarser grains. This is confirmed by the screen analysis (see Table 4.3). Prior to further use, the FBA was hydrated, similar to the FFA. After hydration and drying, compact parts were formed, which were broken in a mortar. Subsequently, this material was sieved through a 200 μm sieve. The sieve residue (55.4 wt%) was not used for the preparation of the experimental samples.

Table 4.3 – Screen analysis of PFA after mixing with water and drying.

fraction / μm	<50	50-100	100-200	200-500	500-1000	>1000
amount / wt.%	0.45	3.25	40.9	35.3	16.8	3.3
fraction / μm	<50	<100	<200	<500	<1000	total
amount / wt.%	0.45	3.7	44.6	79.9	95.7	100

4.2   Samples preparation

The input materials (illite, grog, and fly ash) were first mixed in the required ratio and thoroughly mixed in a dry state. The proportions by weight of the raw materials (in the dried state) and the indications of the individual mixtures to be used further are given in Table 4.4.

Green ceramic bodies were prepared by the wet route. The mixture of input materials was mixed with demineralized water and blended to form a plastic mass suitable for extrusion. The volume of water required depended on the mixture's ability to form a sufficiently plastic mass suitable for extrusion, but in general the water content was approximately 35% by weight. The plasticity of the mass for extrusion was not measured; it had to be estimated based on our experience. If the mass contained too much water, the samples tended to contain many air bubbles. A mass whose water content was too low was difficult to extrude because of the high pressure required.

Extruded cylindrical samples were used for differential thermal analysis (DTA), thermogravimetry (TG), thermodilatometry (TD). They were also used to determine firing weight change, firing dimension change, resonance frequency to calculate Young's modulus, thermal diffusivity (and also thermal conductivity and specific heat capacity), porosity, and pore size distribution.

Table 4.4 – The composition of experimental mixtures in wt.%.

notation	Illite	grog	FFA	PFA	FBA
illite	100				
A0	60	40			
FF10	60	30	10		
FF20	60	20	20		
FF30	60	10	30		
FF40	60	0	40		
PF10	60	30		10	
PF20	60	20		20	
PF30	60	10		30	
PF40	60	0		40	
FB10	60	30			10
FB20	60	20			20
FB30	60	10			30
FB40	60	0			40

Prismatic samples for the determination of flexural strength were made by cutting from a large piece of plastic mass using a thin wire. The prism samples were $10\times10\times100$ mm^3 immediately after the cutting. Such a procedure was chosen for two reasons: 1) the apparatus is adapted to break prismatic samples, and 2) cutting the sample using a thin wire from a large piece of plastic mas that was randomly mixed minimizes the risk of texture formation.

Cube-shaped samples with an edge of ~40 mm for measuring water absorption by boiling were made by molding the wet plastic mass. After drying, each side of the cube was ground by a few millimeters using abrasive paper. A uniform sample surface was achieved by this and the risk of biasing the results due to different water diffusion across the sample surface was eliminated.

After shaping, the samples were dried in the laboratory, covered with a perforated plastic foil to slow down drying. This minimizes the risk of cracking due to rapid drying. Prior to the measurements, samples from the individual mixtures were placed together in a plastic bag to equalize their moisture content. For the same reason, samples from mixtures with the same fly ash were placed together in a hermetically sealed space before DTA and TG measurements.

4.3   Scanning electron microscopy (SEM)

Micrographs of the input materials particles and the microstructure of the fired mixtures were taken by scanning electron microscopy (SEM) at the Materials Research Center of Tallinn University of Technology. A Zeiss EVO 50 series scanning electron microscope was used. The starting materials in the form of powder were applied in small quantities to the double-sided adhesive tape, and the other side of the tape was glued to the sample holder of the microscope. The specimens for fracture surface examination were also attached to the holder by double-sided adhesive tape. Electron removal from the sample surface was ensured by the plasma deposition of the sample surface with palladium and gold atoms and grounding the entire sample holder.

4.4   Particle size distribution

The measurements were carried out using the Annalyset-TE 22 from Fritsch at the Department of Materials Engineering and Chemistry of the Czech Technical University in Prague. The method for determining the particle size distribution on this apparatus is based on the diffraction of the laser beam after passing through the measured material (Frauenhofer diffraction) dispersed in the liquid medium. Ultrasonic vibrations were applied to the medium with the measured material to disrupt the agglomerates. Each measurement was repeated five times, and the resulting values were calculated as the arithmetic mean of the five measurements. The result is either a percentage of particles with a diameter at a selected interval or a cumulative particle size below a certain size [86].

4.5   Specific surface area

Specific surface area (SSA) plays a role in the sintering process, where its reduction (and thus the interface energy reduction) is the driving force of sintering [40]. The size of the SSA also directly affects the amount of physically bound water adsorbed on the particle surface.

SSA was determined by the adsorption method of N_2 molecules on the surfaces of particles. Helium was used as a carrier gas, and the SSA determination was carried out at liquid nitrogen temperature. The Kelvin 1042 sorptometer (Costech Microanalytical SC) at the Laboratory of Inorganic Materials of Tallinn University of Technology was used for measurement. The specific surface is determined by instrument software based on measurements using Brunauer - Emmett - Teller (BET) theory [87].

4.6   Chemical analysis

The chemical composition determines the properties of materials, their structure, color, mineralogical composition, etc. In the building ceramics, the silicon and the oxygen elements are mostly presented. Therefore, it could be classified as silicate ceramic. The other conventional elements are aluminum, calcium, iron, magnesium, and others. The final structure and material properties depend not only on the chemical composition but also on the technology used for the production (temperature, temperature regime, pressure, initial particle arrangement, dispersion, homogeneity, and many others).

The chemical composition in the oxides form was determined by X-ray Fluorescence Spectroscopy (XRF) using a Bruker S4 Pioneer Spectrometer at the Institute of Geology of the University of Technology, Tallinn. An X-ray tube with an Rh anode and a maximum power of 3 kW was used. Approximately 8 g of the sample was mixed with eight drops of a 5% Mowiol solution in distilled water and pressed. The pellets were dried at 105 °C for 2 h. The measured data were evaluated using software supplied by the instrument manufacturer.

4.7 X-Ray diffraction analysis (XRD)

X-ray diffraction analysis is one of the microstructural methods used for the identification of crystalline phases of materials. The phase analyses were performed by powder X-ray diffraction using a diffractometer Bruker D8 Advance with a Cu anticathode ($\lambda \alpha_1 = 1.5406$ Å), accelerating voltage 40 kV, and beam current 40 mA, at Tallinn Technical University. The evaluation of the XRD spectra was performed using Match!© software. The software uses the COD-Inorg REV140301 free mineral database to identify the mineralogical phases. Qualitatively, the individual mineralogical phases are determined by the position of diffraction maxima. A certain knowledge of the chemical composition of the material and the probability of occurrence of the mineral in the material is also helpful. The semiquantitative determination of the number of individual phases is performed in the software by comparing the intensity of individual diffraction maxima (the amorphous phase is not determined). The amount of specific mineral X_{real} was calculated based on knowledge of the amount of quartz determined by DSC analysis (considered as the real amount of quartz) P_k, the amount of quartz determined by semiquantitative XRD diffraction Q_{XRD}, and the amount of the mineral determined by XRD diffraction X_{XRD}

$$X_{\text{real}} = X_{\text{XRD}} \frac{P_k}{Q_{\text{XRD}}} \qquad (4.2)$$

4.8 Differential thermal analysis (DTA), Thermogravimetry (TG), Evolved gas analysis (EGA)

Differential thermal analysis (DTA) is a method that gained its popularity mainly due to its relative simplicity and the quality of obtained results. In the DTA, the temperature difference between studied material and an inert reference sample is measured during the identical thermal treatments (heating or cooling temperature program). It allows determining temperatures at which significant phase changes occur based on information about the release or consumption of heat for these processes. Thermal changes in the sample are detected relative to the inert reference. Two situations can occur (except when there is no reaction):

- Exothermic reaction – heat is released at a specific temperature or time interval. For example, the combustion of organic material, crystallization or change of crystalline modification (from high-temperature to low-temperature modification);

- Endothermic reaction – heat is consumed during the reaction. For example, the release of physically bound water (dehydration), dehydroxylation, dissociation of carbonates, changes in crystalline modification (from low-temperature to high-temperature modification), and melting

The heating rate of the sample has a significant effect on the sensitivity of the method. With lower heating rates of the sample, the results are more reproducible, but this reduces the method's sensitivity because the temperature difference between the reference and test substance is smaller and more easily compensated by heat conduction. When studying ceramic materials, some effects may disappear entirely. If the heating temperature rises too fast, the DTA curve peaks are very pronounced, but their shape and position are less reproducible. In general, the peaks due to rapid reactions disappear. The most commonly used heating rates are $5 - 20$ °C min^{-1}.

Thermogravimetry (TG) is a method used for determining the mass changes of the studied sample as a function of time and temperature. Thermogravimetry is useful:

1. to the correction of mass changes for measuring Young's modulus during heating,

2. to analyze the processes proceeded in the material, under investigation during heating (evaporation, dehydration, thermal decomposition, oxidation, etc.).

The second point also provides the possibility to identify the components of the material and possibly its quantity. However, determining the amount of a given component in the material requires accurate knowledge of the decomposition reaction, i. e. what part of the original component has been released in gaseous form from the material [88].

Evolved gas analysis (EGA) is used to identify gases released during heat treatment. Frequently, the term mass spectroscopy is also used, which better describes the way in which the identification of the individual components of the released gases is realized. The release of material from the sample under investigation, in the form of gases (e.g., H_2O vapor during dehydration or dehydroxylation, CO_2 gas in the combustion of organic residues, SO_2 gas in the decomposition of sulfur compounds, etc.), also results in weight loss and is thus detectable by TG analysis. Based on the temperature at which gas evolution occurred, the process can be determined from the DTA and TG analysis. However, this requires considerable experience with the type of material examined. If unknown material is subjected to the study or even individual processes overlap, EGA is preferably used as an additional analysis of the DTA and TG analysis to identify the processes.

Simultaneous analyses of DTA and TG of compact samples were performed on a reconstructed Derivatograph 1100° [89] up to 1030 °C. The DTA signal represents the temperature difference between the measured and the reference sample. The reference sample's shape and size are close to the measured sample, i. e. a cylinder with a height of ~16 mm and a diameter of 12 – 15 mm. After completing the first measurement, a second measurement (so-called blank) was performed with the same sample already fired, and this measurement was subtracted from the first measurement. In this way, the effect of the apparatus was removed from the resulting curves. The disadvantage is that reversible reactions such as α-β modification of quartz have been eliminated. The relative mass change, Δm, at t was calculated as

$$\Delta m(t) = \frac{m(t) - m_0}{m_0} \, , \qquad (4.3)$$

where $m(t)$ is the mass of the sample at temperature t, and m_0 is the initial mass of the sample. The advantage of the TG analysis of compact samples is a more accurate description of the ceramic body's mass changes during heating. It is caused by the transport of evolved gases through the compact porous sample, contrary to the powder one. For example, closed pores may be formed in the body in which the released gases become trapped or, due to the closer contact of the particles, reactions may occur that would not be observed in the powder sample. The measurements were performed in a static (self-generated) air atmosphere. The temperature program of measurement was set to linear heating at 5 °C min^{-1} to 1030 °C. The disadvantage of this apparatus is the relatively low maximum temperature of 1030 °C.

Simultaneous measurements of DTA, TG, and EGA of powder samples were performed at the Inorganic Materials Laboratory of Tallinn Technical University on a Setaram LabSys 1600 thermoanalyzer coupled to a Pfeiffer Omnistar Mass Spectrometer. The DTA output is the difference of thermoelectric voltages of the thermocouples placed under the measured and reference sample. An empty corundum crucible without a lid was used as a reference. The mass spectrometer was set to detect the released H_2O, CO_2, and SO_2 molecules. A sample with a mass of ~20 mg was placed in a corundum crucible without a lid. The measurements were performed in a dynamic atmosphere of a mixture of 79% Ar + 21% O_2 with a flow rate of 60 ml min^{-1} to a maximum temperature of 1200 °C with a heating rate of 20 °C min^{-1}.

The mass change was also determined at room temperature on samples fired to different temperatures. Five samples were fired successively at temperatures of 100, 200, ..., 1100 °C in the following program: heating 5 °C min^{-1} → soaking time 5 min → cooling 10 °C min^{-1} to room temperature. The samples had a diameter of ~13 mm, a length of 90 – 120 mm, and a weight of between 20 and 26 g (depending on the size and the particular mixture) before the first firing. After firing, the weight of each sample was weighed on a Kern EG 620-3NM digital balance with a readout $d = 0.001$ g. For each sample, the relative weight change over the unfired (raw) sample was calculated, similar to the Equation (4.3). The resulting relative mass change of the test material was calculated as the arithmetic mean of the relative mass changes of the five samples. The uncertainty of the result was characterized by the standard deviation. The uncertainty of weight determination was not included in the resulting uncertainty as it was negligible (less than 0.5%) due to the sensitivity of the scales and the relatively significant weight changes. Mass changes after firing at different temperatures were measured on the same samples, which were used for Young's modulus measurements.

4.9 Volume changes

Processes occurring in the ceramic material during its thermal treatment can result in dimensional changes of the material. The dimensional changes could be monitored at the current temperature (in-situ), thermodilatometry (TD) or after firing to a specific temperature, dilatometry. The advantage of TD over dilatometry is that it is also possible to observe reversible reactions, such as a standard thermal expansion of the material, modification of silica at 573 °C, and others. At the same time, we can accurately identify the temperature at which a specific process (reversible and irreversible) takes place. Knowing the exact development of body dimensions during heating (and especially the rate of its change) is key to optimizing firing. This helps us avoid abrupt volumetric changes, which, due to the size of the body and the rate of heating (or cooling), could lead to body failure. It occurs when the mechanical stresses caused by the different temperature contraction/expansion of adjacent layers of the ceramic body exceed the breaking strength of the material from which the ceramic body is made. Besides, the results of TD measurements can be approached by examining the physical nature of the various processes that cause dimensional changes [84, 90].

TD measurements were performed on a horizontal push-rod dilatometer (detailed description in [91]). The dilatometer's furnace is built by a refractory porous alumina brick with a SiC rod used as heating elements. The sample temperature is measured with a Pt-PtRh10 thermocouple (type S). The heating program is controlled by the adjustable temperature controller CLASIC CLARE 4.0. Dimensional changes are recorded using a linear differential transformer (LVDT sensor), whose output signal is converted to a dimensional change after calibration with a corundum standard. The relative elongation $\varepsilon(t)$ at temperature t was calculated based on the relationship

$$\varepsilon(t) = \frac{L(t) - L_0}{L_0},$$ (4.4)

where $L(t)$ is the length at the current temperature t, and L_0 is the initial length of the sample. The cylindrical samples had an initial length of ~30 mm and a diameter of ~13 mm. The sample bases were ground on a precision grinder to be planar. The pressure force of the 0.33 N was evenly distributed on the sample base using alumina plates, which also protects the alumina pressure rods of the dilatometer from sintering to the sample at high temperatures. The working atmosphere of the dilatometer was static air. The temperature program was set to: heating 5 °C min^{-1} up to 1100 °C → cooling 5 °C min^{-1} to room temperature.

After firing, the dimension changes were determined at room temperature by measuring the sample dimensions before and after heat treatment using a digital caliper with a readout $d = 0.01$ mm. A total of 5 cylindrical samples from the same mixture with a diameter of ~ 13 mm and a length of 90 – 120 mm were fired successively at different temperatures. After firing to the desired temperature, the relative truncation of each sample was determined according to Equation (4.4), where t is the firing temperature, L_0 is the length of the raw sample, and L is the length of the sample fired to t. The resulting relative dimensional change was then calculated as the arithmetic means of the relative dimensional changes of the five samples. The uncertainty of measurement was characterized by the standard deviation from the relative truncation measurements by firing of individual samples. The uncertainty in determining the length of the individual samples was not taken into account since it was negligible (less than 0.5%) due to the caliper's sensitivity and the relatively significant dimensional changes. The firing temperature program was: heating 5 °C min^{-1} → soaking time 5 min at maximum temperature → cooling 10 °C min^{-1} to room temperature. Changes in the dimensions after firing to different temperatures were measured on the same samples at the same time as Young's modulus.

4.10 Bulk density

The bulk density (ρ) was determined after firing at different temperatures and during firing at the current temperature. Bulk density after firing is essential from an application point of view, and generally, the smaller value of bulk density is the better. The reduction of ρ should not reduce the mechanical strength below the technical standards, mainly if the material was used for the production of bricks, slabs, tiles, etc. Lighter materials, while maintaining mechanical strength, allow the construction of larger structures. With decreasing ρ the thermal conductivity usually decreases, which is another positive for envelope masonry materials. Knowledge of volume density development during heating is essential for optimizing the firing regime, as it simultaneously reflects changes in weight and dimensions.

After firing, measurements were carried out on the same samples where the change in weight and dimensions by firing was determined. For each of the five fired samples, the bulk density was determined as the proportion of its weight and volume. Then the resulting ρ was calculated as the arithmetic mean of ρ of these five samples. The uncertainty of the results was characterized by the standard deviation from the values of 5 samples.

During the measurement, the ρ at temperature t was calculated based on the results of thermogravimetry (Equation (4.3)), thermodilatometry (Equation (4.4)) and the initial bulk density ρ_0 according to the relation

$$\rho(t) = \frac{1 + \Delta m(t)}{\left(1 + \varepsilon(t)\right)^3}\, \rho_0. \tag{4.5}$$

4.11 Open porosity and pore size distribution

To better understand pores in ceramic materials, it is useful to know the open porosity and pore size distribution behavior. Porosity is one of material properties, which can have a positive effect in the correct application of porous materials. A larger porosity is desirable, for example, for materials with a low density, thermal insulation, for a purpose of water diversion, filtration, etc. Negative effects of porosity are brittleness, a decrease of mechanical strength, or unwanted water absorption [17]. In ceramic materials, voids (pores) occur naturally in consequence of the formation process of raw bodies from powders. Their proportion depends on a formation process. The usage of methods of body formation at low pressure usually leads to higher porosity [17]. The firing process of ceramic materials generally decreases porosity, mainly, if sintering runs through a viscous flow, when pores are closing by melt [9]. A decrease in pores volume causes a shrinkage of the ceramic body during sintering. On the other hand, burning out fillers in a ceramic body caused an additional porosity. Porosity can be also increased by a release of gaseous compounds for example during decomposition of carbonates. Total porosity after a firing is a function of initial porosity and a function of the size of pores. Small pores have higher thermal "reactivity" than big pores. They are removed from a structure easily by a sintering [17]. The size of pores, after a firing, depends on a grain-size distribution, on the shape of initial powders, and on the content of various additives [17]. For building materials, the size pore distribution is important from the point of view of frost resistance. In general, it is valid that bigger pores promote better frost resistance [18].

Measurements of pore size distribution were performed by the device Quantachrome PoreMaster GT®. The pore size distribution is determined by the mercury injection method into pores of the measured sample during increasing pressure. A minimal diameter of pores r which are filled with mercury in the pressure P is given by the Washburn equation [19]

$$r = -\frac{2\gamma \cos \phi}{P}, \tag{4.6}$$

where γ is the surface tension of mercury and ϕ is contact angle between mercury and sample. Apparatus gradually increases pressure from 0 up to 400 MPa. The surface tension of mercury is 0.48 N m^{-1}, and the estimated contact angle was $140°$. The device can measure pores with radius from 1 mm to 3.5 nm. Samples were made by mechanical pressing, and one piece, which was able to put in a sample holder with dimension $\varnothing 8 \times 15$ mm^2, was selected for the measurement of porosity. The result of a measurement is a relation between volume saturation and the diameter of pores.

The open porosity was calculated with the help of the experimentally determined bulk density and matrix density. The bulk density was obtained from the volume and mass of the cylindrical samples. The matrix density was measured by means of helium pycnometry (Pycnomatic ATC, Thermo Fisher Scientific).

4.12 Thermal diffusivity, thermal conductivity, and specific heat capacity

Material parameters such as thermal conductivity λ [W m^{-1}K^{-1}], thermal diffusivity a [m^{2}s^{-1}], and specific heat capacity c_p [J kg^{-1}K^{-1}] are essential for building materials in term of their application, thermal insulation, and a design of construction. While thermal diffusivity characterizes the ability of a material to balance thermal differences in different parts, the thermal conductivity indicates how much thermal energy can pass through a material per time in a given thermal gradient. For building ceramics, thermal conductivity is very important. For practical applications (for example, bricks for enclosure walls), the producers try to manufacture bricks with a thermal conductivity as low as possible. This task is often achieved with an increase in the porosity of the ceramic body. On the other hand, an increase in porosity decreases mechanical strength in general. Hence, it is necessary to find a compromise between these two properties.

Specific heat capacity defines how much thermal energy must receive a body with unit mass to increase its temperature by one unit in temperature. The knowledge of the specific heat capacity of building ceramics has significance in terms of the thermal comfort of buildings. For example, if buildings are built from materials with a high heat capacity, to reach thermal comfort (required temperature) of the specific place during the heating season takes a longer time. Moreover, much thermal energy is spent. On the other hand, the temperature in the specific place decreases slowly when the heating is going to shut down. The buildings with a high heat capacity have advantages in the summer season when a building construction absorbs solar radiation during the day, whereby the temperature is changing very slowly, and the accumulated heat is gradually releasing during the night. In the last decades, the materials with a significant ability to absorb or release a large amount of thermal energy at relatively narrow temperature interval are intensively developed. The ability of these materials to bond a large amount of thermal energy is reached mainly based on the change of a solid phase of matter into a liquid phase and backward or based on the reversible change of a structure at a certain temperature. The knowledge of the specific heat capacity has also significant in terms of a firing optimization of ceramic materials.

Thermal diffusivity, thermal conductivity, and specific heat capacity were determined using apparatus LFA 427 LaserFlash from the Netzsch company at which the basic quantity measured by this device is thermal diffusivity. The measurement is based on applying a very short (0.3 – 1.2 ms) laser pulse with a wavelength of 1064 nm on the bottom face of the measured sample. Consequently, the increase in a temperature on the upper face of the sample as the dependence of time is measured. From the character of a temperature increase on an upper face of a sample (see Fig. 4.1) thermal diffusivity of the studied sample is determined. Several models can be used for the calculation of thermal diffusivity. These models consider several effects, such as the loss of heat from sides of a sample, the heat transfer by radiation inside a sample, or the correction of a laser pulse width. Then, a value of the thermal diffusivity a is determined by a non-linear optimization during the solving of differential equations of heat conduction corresponding to the given models.

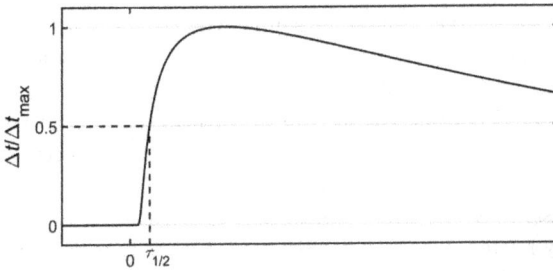

Fig. 4.1 – Typical behavior of temperature vs. time during the measurement of thermal diffusivity using a flash method.

The apparatus LFA 427 LaserFlash uses a ruby laser excited by a xenon lamp for forming a laser pulse with a wavelength of 1064 nm. A laser pulse is transported to the dispersion glass (lens) through an optical cable. The temperature on the measuring (upper) side of a sample is measured using an InSb infrared detector cooled by liquid nitrogen. The intensity of laser pulse is possible to regulate by a voltage of the xenon lamp. The voltage was set up on the value of 510 V (this is about 30 V more than the lower adjustable value of the voltage in the device). The pulse length was set up on the value of 0.3 ms (this is the lowest adjustable value).

The measurement of the specific heat capacity of the studied sample requires an additional measurement of reference material with known specific heat capacity and density. This method's basis is in applying a laser pulse with the same parameters of measured and reference sample. In this way, the same amount of heat is provided to both samples. Next, the limit adiabatic temperature is calculated for both measurements. The specific heat capacity c_p of the measured sample is calculated according to the formula

$$c_p = \frac{m_R \, c_{pR} \, \Delta T'_\infty}{m \Delta T_\infty},$$

(4.7)

where m_R is a mass of the reference sample (in software, a mass of a sample is calculated from its bulk density and dimensions), c_{pR} is a specific heat capacity of the reference sample, $\Delta T'_\infty$ is an adiabatic temperature of the reference sample when a given amount of thermal energy is reached, m is a mass of the studied sample, and ΔT_∞ is the adiabatic temperature of studied sample when the same amount of thermal energy as for reference sample is reached.

Thermal conductivity of the material λ is calculated as

$$\lambda = a\rho c_p,$$

(4.8)

where ρ is a bulk density of the studied material.

Samples for the measurement of thermal diffusivity were cut from an extruded dried cylinder with a required diameter 12.5 mm and length of about 2.8 mm. Consequently, this sample was ground on the final thickness of about 2.5 mm so that the faces of a sample were plane-parallel. This was possible using the template. As reference material for determining specific heat capacity, the glass-ceramic material Pyroceram with a diameter of 12.5 mm and a thickness of 2.5 mm was used. This material was used because its value of thermal diffusivity is ~1.8 mm^2 s^{-1}, and this is the closest value of thermal diffusivity (from the supply reference materials) to the studied samples (0.3 – 0.6 mm^2 s^{-1}). Before each measurement, the layer of graphite (using spray) was applied on both studied and reference samples in order to provide the repeatability of surface properties which is very important in respect of the character of measurement (it is based on the absorption of radiation). Besides, the graphite has a high thermal diffusivity, and because of that, it assists in distributing a temperature in the whole bottom face of a sample after an application of laser pulse. Thereby, the experimental conditions are closer to the theoretical one. For the calculation of thermal diffusivity, the model contains heat transport by conduction as well as heat transfer by radiation inside of a sample and a correction of the width of a laser pulse.

Three samples from each mixture were prepared. The samples were fired at temperatures from 100 °C up to 1100 °C with a heating rate of 5 °C min^{-1} and a soaking time at maximal temperature 5 min. Then, the samples were cooled down to room temperature with a cooling rate of 10 °C min^{-1}. After firing, the dimension of samples was determined, and then their values were added to the apparatus software. Each dimension was measured 10 times by caliper with a resolution $d = 0.01$ mm. The combined uncertainty of a thickness determination did not exceed more than 1 %. After a dimension measurement, two graphite layers were applied on each side of a sample. Graphite layer was also applied on a reference sample (reference sample was not fired at different temperatures). Thermal diffusivity was determined at temperature (30 ± 2) °C in the air atmosphere and was measured three times in the interval of 4 min for each sample. The final value of thermal diffusivity was calculated as the arithmetical average of these three measurements. The bulk density of studied samples was determined from a mass and dimension of samples which were used for the measurement of Young's modulus (the dimension of samples was $\varnothing 12 \times 110$ mm^2).

Thermal diffusivity was measured at room temperature as well as during heating by the apparatus Netzsch LFA 427 LaserFlash. The thermal regime was as follows: a linear heating with a rate of 5 °C min^{-1} up to selected temperature, then during measurements, isothermal heating was performed. The selected temperatures were increased by 50 °C from 50 °C up to 1100 °C. The measurements were performed after temperature stabilization (the maximal deviation of a temperature in the interval ±4 °C; the maximal change of a temperature 1 °C min^{-1}; the stabilization of signal of infrared sensor 0.2 V s^{-1}). After stabilization, two measurements were carried out for the determination of thermal diffusivity at a selected temperature. The value of thermal diffusivity at the selected temperature was calculated as arithmetic average from these two measurements. The average heating rate was 3.3 °C min^{-1} at selected configuration. The experiment was performed in the dynamic N$_2$ atmosphere with a rate of 100 ml min^{-1}. The reason was to protect a graphite layer on a sample from thermal oxidation at temperatures above 700 °C. The correction of a thermal expansion during a heating was done using thermodilatomery measurements. These measurements were carried out by a dilatometer Netzsch DIL 402 C with a heating rate of 3.3 °C min^{-1}.

4.13 Young's modulus

Young's modulus E is related to the elastic properties of materials. This quantity is very important for the design of machines, devices, or buildings. It is a degree of expansion or contraction of a material under the effect of a direct tensile or pressure stress in the direction of this stress [20]. If this stress is relatively small (i.e., it is not adequate enough to register a permanent plastic deformation), the relation between stress and relative expansion is given by Hooke's law (a one-dimensional case) [20]

$$\sigma = E\varepsilon, \tag{4.9}$$

where $\sigma = F/S$ is stress given by a force F acting on a surface of a cross-section S, which is perpendicular to the force direction. The relative expansion (strain) $\varepsilon = \Delta l/l_0$ is given by the ratio of an expansion Δl and an initial length l_0 during acting of stress σ.

The values of Young's modulus can be used for a qualitative estimation of mechanical strength. According to the Griffith theory, the tensile strength τ_t of brittle materials (such as ceramics) can be expressed as [21]

$$\tau_t = C\sqrt{E}. \tag{4.10}$$

It is visible that the tensile strength increases with an increase of Young's modulus. An increase or decrease of the tensile strength of a material can be deduced based on a relation of Young's modulus as a function of a temperature during a firing or after a firing at different temperatures. The advantage is that methods for measuring Young's modulus are non-destructive in contrast to the direct measurement of mechanical strength. However, it is necessary to have in view that the parameter C in Equation (4.10) is directly proportional to surface energy and indirectly proportional to the length of the biggest rapture. Therefore, if these effects are significant, the tensile strength can decrease even Young's modulus increases.

The methods for determining Young's modulus can be divided into static and dynamic, according to the time dependence of an acting force. The simplest static method is a direct application of Hooke's law, Equation (4.9), when a force affects a studied sample, and a strain is measured. However, this approach is appropriate only for materials with relatively small Young's modulus. If the values of Young's modulus are high, it is necessary to produce an extensive force to reach a measurable strain. Also, a problem with the fixation of a sample exists. Additionally, for brittle materials, a brittle fracture occurs earlier than measurable deviation is achieved.

Dynamic methods are the alternative to static methods. They are based on the interrelation between Young's modulus and the resonant frequency of mechanical vibrations. Details of a vibration problem of bodies with simple shapes can be found in [22, 23]. The correct determination of a resonant frequency is the experimental problem. In this research, the impulse excitation technique (IET) was used. A sample was fixed in nodal points of the fundamental mode of vibration. This means it was fixed in two points with a distance of 0.224 times of a sample length from both ends. The fixation of a sample in nodal points is important because a vibration should not be damped with external forces. Thereafter, a mechanical vibration was excited by the impactor (a metal ball on the end of an elastic stick) with a hit in the middle of a sample (i.e., in antinode). This was recorded using an electret microphone, and through a preamplifier, a signal was brought into a computer (sampling frequency of 44 kHz). Then, its frequency spectrum was calculated using Fourier analysis. The example of the recorded signal and calculated frequency spectrum is in Fig. 4.2. The software for recording signals and its evaluation was created in the Matlab software.

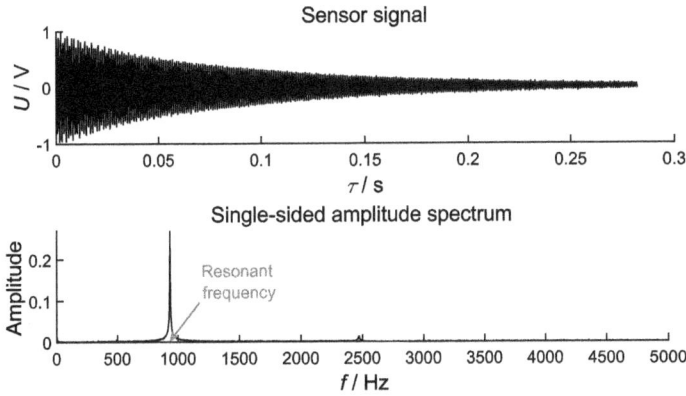

Fig. 4.2 – The example of recorded signal after an excitation of a vibration and its frequency spectrum.

Young's modulus is calculated as follows [24]

$$E = \left(K\frac{l^2 f}{d}\right)^2 T\rho,$$ (4.11)

where $K = 1.12334$ (for a cylindrical cross-section of the sample), l is the length of the sample, d is the diameter of the sample, f is a resonant frequency of fundamental mode of flexural vibration, and ρ is density. The correction coefficient T is needed if the ratio $l/d < 20$. It is calculated according to the norm ASTM C 1259-01 [24]

$$T = 1 + A(1 + 0{,}0752\mu + 0{,}8109\mu^2)\left(\frac{d}{l}\right)^2 - B\left(\frac{d}{l}\right)^4$$
$$- \frac{C(1 + 0{,}2023\mu + 2{,}173\mu^2)\left(\frac{d}{l}\right)^4}{1 + D(1 + 0{,}1408\mu + 1{,}536\mu^2)\left(\frac{d}{l}\right)^2},$$ (4.12)

where μ is the Poisson ratio. For isotropic materials, it can have values from 0 to 0.5. The value of $\mu = 0.3$ was added into Equation (4.12) (typical for ceramic materials on the base of illite) [25].

Young's modulus was determined after firing at different temperatures using five samples from each mixture. The samples were fired at temperatures of 100, 200, ..., 1100 °C. The dimensions, mass, and resonant frequency of the fundamental mode of flexural vibration were measured after cooling at room temperature. Thereafter, Young's modulus was calculated from these measured values according to Equation (4.11) for each sample. The final value of Young's modulus for each mixture after a firing was determined as an arithmetic average of five samples. The uncertainty of Young's modulus determination is characterized by a standard deviation of five values of Young's modulus. The relative expanded uncertainty of Young's modulus is ~1 % [26].

The samples were fired in the electric furnace in a static air atmosphere. The regime of a firing was: linear heating with a rate of 5 °C min^{-1} → isothermal heating at maximal temperature for 5 min → linear cooling with a rate of 10 °C min^{-1} to room temperature.

Youngs' modulus during firing was determined on cylindrical samples with a diameter of $\varnothing$12 mm and a length of 110 mm. The length of a sample was chosen so that it was fixed in nodal points of the fundamental mode of flexural vibration. An electrodynamical impactor with a steel core was used for an exciting flexural vibration. This impactor was coupled with an alumina rod, which extended to the working space of a furnace (a temperature up to 1200 °C). Alumina rod hit a sample approximately in the middle in set intervals. For the detection of a sample vibration, the apparatus uses an electret microphone. This microphone cannot be placed close to a sample (due to a high temperature). Hence the alumina sound pipe was used for the transmission of an acoustic signal. A signal from the microphone was transported in the microphone input of a computer through the acoustic preamplifier. In computer, the signal was recorded with a sampling frequency of 44 kHz, and then a frequency spectrum was calculated using the fast Fourier transformation. The resonant frequency is a frequency with the maximal amplitude. Often it happens in the spectrum that parasitic frequencies with a higher amplitude as a resonant frequency occur. The software allows specifying a narrow interval of frequencies where a resonant frequency is seeking to not evaluate parasitic frequencies as a resonant. The boundaries of this range are shifted automatically with a shift of the resonant frequency. The software manages whole measurements. It activates an electrodynamical impactor in chosen intervals, records signal, and measures an actual temperature in the furnace (Pt-PtRh10 thermocouple, type S). The scheme of the measuring apparatus is in Fig. 4.3, and its detailed description is in [27]. For a calculation of Young's modulus was used Equation (4.11) with a correction of dimension and mass changes of a sample from a thermodilatometric and thermogravimetric data [27]. The relative expanded uncertainty of Young's modulus determination at higher temperatures is ~1.1 % [27].

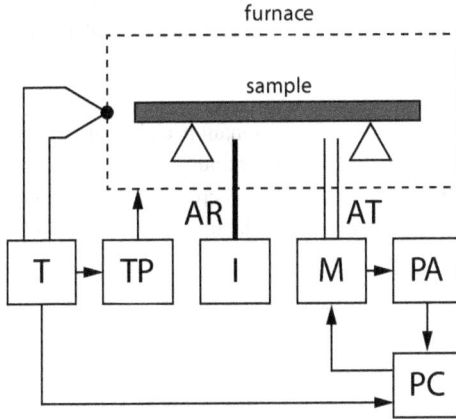

Fig. 4.3 – The scheme of apparatus for the measurement of a resonant frequency by impulse excitation method at higher temperatures. TP – temperature programmer, AR – alumina rod, AT – alumina tube, I – impactor, M – microphone, PA – preamplifier, PC – computer, T – type S thermocouple.

4.14 Internal friction

The internal friction characterizes the non-elastic properties of materials. After excitation of mechanical vibration, its amplitude starts to decrease with time exponentially (see Fig. 4.2). The speed of this decrease is not always the same, but it depends on the material. The reason to decrease the amplitude of vibration is that the energy of mechanical vibration is changing to another form of energy (for example, acoustic or thermal). Without consideration of interaction with surroundings (i.e., a sample levitates in a vacuum) energy of a mechanical vibration can transform only to thermal energy (it is internal damping). The transformation occurs due to a dispersion of vibration energy on defects of a material structure (grains boundaries, impurities, microcracks, etc.), due to thermoelastic effects caused by temperature gradients, or due to dislocation movements (mainly for metals) [28].

The internal friction can be quantified by several quantities with the correlation between them [29, 30]. In this research, the logarithmic decrement δ is used to quantify internal friction [30]. It is defined as a natural logarithm of two successive amplitudes of vibration, and it was determined from the shape of the resonant curve. For this purpose, it is necessary to find the width of a resonant peak in half of its height Δf (see Fig. 4.4). The resonant curve was derived from the result of fast Fourier transformation of the record, which was obtained after the excitation of a vibration. The points in the half of the maximal height of resonant peak (f_1 and f_2 in Fig 4.4) were obtained by linear interpolation between two nearest points. Then, the logarithmic decrement δ was calculated according to formula

$$\delta = \frac{\pi \Delta f}{\sqrt{3}}.$$
(4.13)

It was determined from the width of the resonant peak because beats often generate during vibration of a cylindrical sample, and then a reliable determination of a vibration amplitude is not possible (see [30]). The logarithmic decrement was determined together with the measurement of a resonant frequency of a flexural vibration on samples after and during a firing as well.

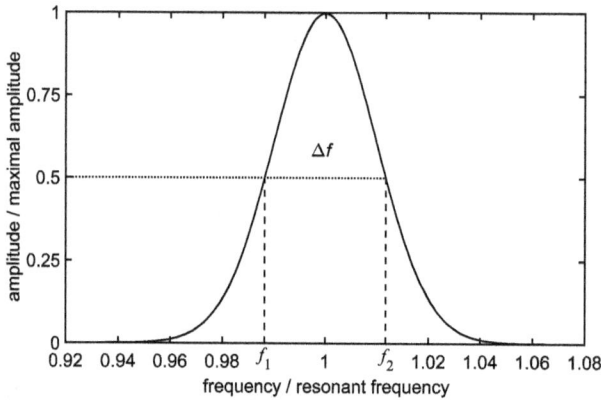

Fig. 4.4 – Determination of logarithmic decrement from a width of the resonant peak.

4.15 Mechanical strength

One of the essential material parameters of building ceramics is mechanical strength. It directly determines their usability for practical applications and the structure's durability. Depending on the different purposes, different requirements are placed on it. If the mechanical strength of the material is sufficiently high, smaller and lighter products can be produced. On the other hand, achieving sufficient mechanical strength requires increased production investment in higher firing temperatures, adding various admixtures, etc. Often a compromise must be sought between sufficient mechanical strength and other material parameters. For example, an increase in porosity decreases thermal conductivity, which is advantageous for building envelope masonry elements. However, as porosity increases, compressive strength decreases, and, therefore, the porosity of materials can only be increased to a certain extent by an appropriate mix design. Mechanical strength is not considered as physical but as an engineering quantity. Its values depend on the testing conditions. Therefore, only the mechanical strengths of materials measured under the same conditions can be compared with each other. In order to make such comparisons possible within the framework of industrial production, industrial standards are drawn up by the standardization bodies to define precisely the conditions for testing the mechanical strength of materials. Mechanical strength can be measured in tension, compression, bending, shear, or torsion. Flexural strength is of particular importance for tiles. For bricks, mechanical compressive strength is more important. The mechanical compressive strength is typically four to eight times higher than the mechanical flexural strength [92, 93].

In this work, the mechanical strength (flexural strength) R of materials was measured after firing to different temperatures. The measurement was carried out by the three-point bending method [94] at room temperature on the apparatus of our own design. Due to the relatively large scatter in the results, ten specimens were used to determine each value. The final value was calculated as the arithmetic mean R of the ten samples. Samples whose R was outside the interval of twice the standard deviation of the sample set were excluded. Since this method is destructive and sample preparation laborious, the R of each mixture was determined after firing to only four temperatures: 120, 800, 1000, and 1100 °C. The temperature of 120 °C was chosen to assess the R of the raw bodies after drying, which is essential for handling ceramic materials before firing. Higher temperatures were selected to see how the increased firing temperature affects the R values. to the desired temperature at a rate of 5 °C min^{-1} → isothermal residence at this temperature for 5 min → cooling at a rate of 10 °C min^{-1} to room temperature.

The span L of the supports with circular cross-section on which the specimens were placed was (74.50 ± 0.10) mm. The clamping force was applied to the specimen at its center by a mechanism that distributed this force evenly across the width of the specimen (using a free-jointed connection). The specimens were prisms with a square cross-section of about 6.5 × 6.5 mm^2. The length of the samples was approximately 80 mm. R was calculated as [94]

$$R = \frac{3FL}{2bh^2},$$
(4.14)

where F is the force at which the specimen broke, b is the width of the specimen, and h is its thickness. The rate of increase of the loading force was (4.80 ± 0.05) N s^{-1}. Therefore, the estimated relative measurement uncertainty R of a single specimen is 5%.

The flexural strength of dry-pressed ceramic tiles should be more than 2.0 MPa in the raw state. For smaller ceramic tiles, a flexural strength value of 1.5 MPa is sufficient [12]. In [14], the authors state that the flexural strength of the raw body should be at least 0.7 MPa.

According to BS EN 14411, ceramic tiles produced by extruding should meet the requirements in Table 4.5.

Table 4.5: Requirements for extruded ceramic tiles according to BS EN 14411.

	Class				
	AI_a	AI_b	AII_a	AII_b	$AIII$
Mechanical strength / MPa	>28	>23	>20	>17.5	>8
Water absorption / %	<0.5	0.5 - 3	3 - 6	6 - 10	>10

5. Results and discussion

This chapter is devoted to the obtained results, their interpretation and discussion. The first subchapters deal with the results of analyzes of input materials used for the production of experimental samples. The study of individual mixtures follows in order to see the influence of the power plant ash mixture on the firing process and the final properties of illite-based ceramic materials. The results concerning the experimental mixtures are divided into subchapters according to the individual analyzes. The intention is the use of the investigated mixtures as a material for the production of building ceramics, and therefore the results will be interpreted with regard to this assumption.

5.1 The study of input materials

Before analyzing experimental mixtures, it is useful to know the properties of the precursors for their preparation. Based on this knowledge, the origin of the observed processes can be identified later in the analysis of the mixtures. Five input raw materials used for the preparation of experimental mixtures are analyzed:

Illite – strictly speaking, it is illitic clay, but since it contains a large amount of mineral illite (above 70%), it is hereinafter simply referred to as "illite";

Grog – used to simulate temper. It was made by firing powder illite at 1100 ° C with a shelf life of 90 min;

FFA – fly ash from fluidized bed combustion boiler, power plant Nováky, Slovakia. Only the fraction below 200 μm hydrated ash is used;

PFA – fly ash from pulverized coal combustion boiler, power plant Nováky, Slovakia. Only the fraction below 200 μm hydrated ash is used;

FBA – bottom ash from fluidized bed combustion boiler. Only the fraction below 200 μm hydrated ash is used.

The description of the preparation of input materials is in section 4.1.

5.1.1 The morphology of particles

Morphology, i.e. the shape of the particles, depends on the properties of the material from which they are formed and on the conditions under which they are formed. Particle morphology is related to the properties of powder materials, such as bulk density, flowability, surface area, etc. [95]. Here, morphology is studied based on scanning electron microscopy (SEM) micrographs.

The morphology of the illitic clay particles is shown in Fig. 5.1. The predominant part consists of illite crystals which have a platelet shape. They are up to 3 μm in size and are combined into aggregates to form larger particles. The plate-like shape of the illite crystals causes their preferential orientation during shaping, thus forming a technological texture. The properties of illite crystals are anisotropic. If a texture is created in the body, the anisotropic properties of the individual crystals also affect the anisotropy of the whole body [24].

Fig. 5.1– SEM of illite used for the preparation of samples.

Illite was annealed at 1100 °C with 90 min holding time was used to simulate the traditional grog. As can be seen from Fig. 5.2, platelet-shaped illite particles were not preserved. At high temperatures, they melted and bonded. The driving force behind the process of bonding the molten remnants of illite particles was the reduction of surface energy. The fact that it is a melt can be seen in Fig. 5.3 according to the smooth shape of the surface and rounded edges. The porosity is small, and many particles with a diameter above 10 μm have been formed.

Fig 5.2 – SEM of thermally treated illite (grog) used for the preparation of samples.

As can be seen from Fig. 5.3, FFA is made up of small particles of irregular shape, with a large proportion of pores. They could be characterized as "spongy." The degree of sintering of the particles during coal combustion is small due to the low combustion temperature (~850 °C). The material contains quite a few small particles that coat the surface of the larger ones.

PFA contains relatively large, solid particles (see Fig. 5.4). They are predominantly glassy, which will be clear later from the results of the mineralogical composition. The particle shape is rounded, oblong with a rough surface. Smaller (~10 µm) spherical particles called cenospheres are also observed. They are hollow and other smaller particles may be enclosed inside of them. They are formed by condensation of gaseous SiO_2 and Al_2O_3 during rapid cooling in the chimneys [96].

Fig. 5.3 – SEM of FFA used for the preparation of samples.

The SEM micrograph of the FBA fly ash (see Fig. 5.5) is not of very high quality due to problems with electron removal from the particle surface during imaging. However, it can be seen that the FBA contains predominantly particles with a diameter just below 200 µm. Only a small amount of finer particles is observed. The large particles are most probably quartz grains, as shown by the mineralogical analysis (see Table 5.5). The particles have an elongated shape with an irregular surface and places with sharp edges.

Fig. 5.4 – SEM of PFA used for the preparation of samples.

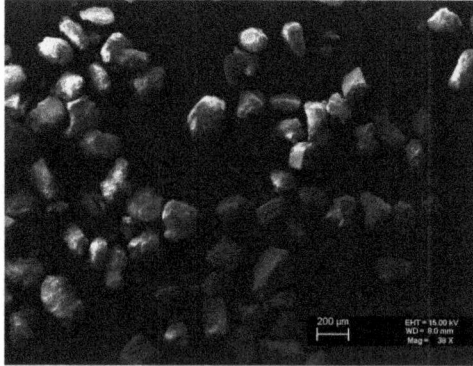

Fig. 5.5 – SEM of FBA used for the preparation of samples.

5.1.2   *Particle size distribution and specific surface area*

The particle size distribution of the materials used to make the samples is shown in Fig. 5.6. Each curve is the average of five measurements, although they did not differ significantly from each other. Illite has two maxima, one at 0.5 µm and the other at 30 µm. Particles below 2 µm are considered to be carriers of the plasticity of ceramic mixtures. Illite contains most of these particles among all input materials. Grog is practically free of particles below 2 µm and its particle size distribution has two maxima, one at 30 µm and the other at ~60 µm. This means that in the process of preparing grog from the illite, the smallest particles were sintered. This is consistent with the SEM observations (see Fig. 5.3). FFA contains particles with a diameter above 10 µm and its particle size distribution is bimodal with one maximum at 30 µm and the other at 100 µm. PFA consists of particles in the range of equivalent spherical diameters from 30 µm to 200 µm, which are represented in approximately equal proportions. Only the fraction below 200 μm was separated from the original PFA ash by sieving so that the part above 200 μm in Fig. 5.6 is a consequence of the inaccuracy of the measuring apparatus. FBA is the coarsest of all input materials because it contains practically only particles close to 200 μm.

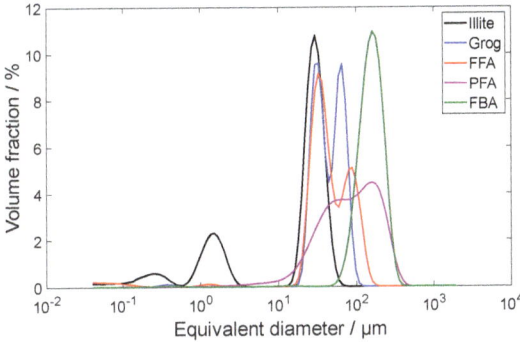

Fig. 5.6 – Particle size distribution of input materials used for the preparation of samples.

The specific surface area (SSA) of the input raw materials is summarized in Table 5.1. If the particles do not contain pores, the larger the proportion of small particles, the larger the specific surface area. The largest specific surface had illite followed by FFA. Significantly smaller specific surfaces had grog, PFA, and FBA. Considering that the grog contains much smaller particles than PFA and FBA (see Fig. 5.6), it can be said that PFA and FBA contain porous particles. The illite particles were sintered, melted, and their surfaces smoothed (see Fig. 5.2), which is the cause of their small SSA.

Table 5.1 – Specific surface area of input materials in $m^2\ g^{-1}$.

Illite	Grog	FFA	PFA	FBA
29.1	1.2	24.6	3.7	3.6

5.1.3   Chemical composition

The results of the chemical analysis performed by the XRF method applied to the input materials are shown in Table 5.2. For clarity, Fig. 5.7a shows the ternary diagram of the SiO_2-Al_2O_3-CaO system (CAS diagram) of the input raw materials used. It can be seen that all ashes had similar SiO_2 content to illite. This indicates good compatibility of the ashes with the basic plastic component (illite). Significant variability in Al_2O_3 content is observed among individual ashes.

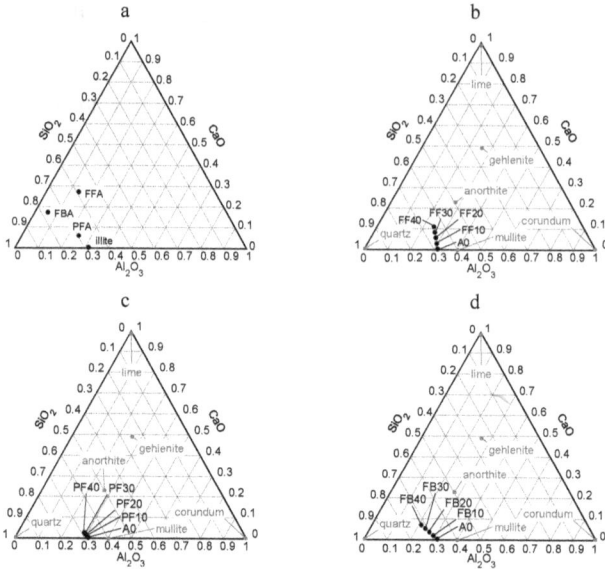

Fig. 5.7 – Ternary diagrams of CAS systems of input materials (a), mixtures with FFA (b), mixtures with PFA (c), mixtures with FBA (d).

The chemical composition of grog is the same as the chemical composition of raw illite. The basis of this assumption is the fact that during firing, mostly water molecules, which are not included in the chemical analysis, are released from the illite.

The CAS diagrams of the individual mixtures are shown in Fig. 5.7b-d. The position of reference sample A0 in the CAS system is identical to the raw illite. The diagrams also indicate the areas of occurrence of some minerals from the CAS system. PFA is closest to illite in its chemical composition (see Fig. 5.7a). FBA contains an increased amount of SiO_2 at the expense of Al_2O_3. FFA contains a lot of CaO, which is the furthest away from the composition of illite.

The chemical composition of mixtures with FFA approach the stoichiometric composition of the anorthite (see Fig. 5.7b) so that its formation can be expected during firing. The PFA (see Fig. 5.7c) has the chemical composition closest to the illite, hence the position of the mixtures with PFA in the ternary diagram does not change much. FBA contains a lot of silica in the form of quartz, which is stable within the temperatures used in this research so that the part of SiO_2 bound in the quartz is not expected to participate in the formation of glass or new minerals.

CAS diagrams serve only for approximate orientation because they do not contain other important oxides such as K_2O, Na_2O, Fe_2O_3, MgO, which can significantly affect the development of microstructure during heat treatment, even in small amounts. For example, illite contains up to 7.63 wt.% of K_2O, which aids in the formation of the melt and significantly reduces the viscosity of the glassy phase. PFA contains 10.61 wt.% Fe_2O_3, which causes a red to dark brown color of the ceramic bodies, depending on the firing temperature. It is also considered a nucleating agent that aids in crystallization. PFA also contains quite a lot of K_2O (2.00 wt.%) and Na_2O (0.80 wt.%), which help to form an amorphous phase.

Table 5.2 – Chemical composition of input materials in the equivalent oxide form as obtained by XRF analysis (in wt.%).

	SiO_2	Al_2O_3	CaO	Fe_2O_3	K_2O	MgO	TiO_2	Na_2O	SO_3	sum
Illite	57.40	26.43	0.28	1.40	7.63	1.34	0.11	0.03	0.03	94.62
FFA	42.18	9.88	19.44	4.75	0.99	2.84	0.45	0.55	8.32	88.85
PFA	55.90	19.66	4.80	10.61	2.00	1.89	0.59	0.80	0.47	95.92
FBA	63.10	4.54	14.20	2.25	0.75	1.13	0.29	0.54	8.15	94.41

5.1.3.1 The content of heavy metals

Representations of heavy metals in fly ash from powerplant Nováky are in Table 5.3. The analysis was prepared on an annual backfill from 2014 to the order of powerplant Nováky. The coal from the Handlová mine, which is burned in Nováky, is characterized by a relatively high arsenic content, which is also reflected in the ash produced by burning this coal. According to Erol et al. [75], heavy metals in sintered material made from power plant fly ash are successfully immobilized, so they do not pose a risk associated with their release from the material. In the case of the successful application of fly ash to the production of building ceramics in terms of mechanical and thermal properties, it will be necessary to examine in more detail the risks associated with the possible release of heavy metals.

Table 5.3 – The content of heavy metals in ashes from powerplant Nováky in mg/kg of ash. Source of an analysis: powerplant Nováky 2014.

Element	FFA	PFA	FBA
Hg	0.16	0.059	0.014
As	705	569	1470
Cd	<1.5	<1.5	<1.5
Pb	17.5	14.6	<9.0
Ni	53.7	35.4	16.2

5.1.4   Mineral composition

The basic experimental method for identifying the phase composition of ceramic materials is XRD analysis. One of the most common mineralogical components of traditional ceramics is quartz. It shows a reversible modification change at 573 °C. This fact can be advantageously used to determine its amount in the material.

5.1.4.1 Determination of quartz content in input materials by DSC method

Based on the change in quartz enthalpy during the α-β modification transition, the amount of quartz in the input materials was determined after annealing at 900 °C. The annealing was performed in order to eliminate irreversible processes that could overlap with the area of quartz modification. The annealing temperature is not high enough to form additional quartz and thus to influence the results. The measurement results evaluated using the Netzsch Proteus v.6.0.0 firmware are shown in Fig. 5.8. The peak area is directly proportional to the amount of quartz in the material, with 100% quartz corresponding to an area of 7.51 J g^{-1} (as determined by measuring pure quartz sand). The calculated amounts of quartz in the individual input materials are summarized in Table 5.4. The estimated relative uncertainty in determining the amount of quartz is 10%. In the case of illite, two substantially different values are given: 13 wt.% and 0.86 wt.%. This is due to the fact that when measuring illite, a superposition of two peaks was recorded in the temperature region of the α-β modification of the quartz (see the inserted graph in Fig. 5.8). Both of these peaks are reversible. The exact cause of this phenomenon is not yet clear. The more probable value of the amount of quartz is 13 wt.%, which is judged from its significant effect on thermal expansion, modulus of elasticity, and also corresponds to the results of XRD analysis. A similar phenomenon is also observed in grog, i.e. illite fired at 1100 °C with 90 min isotherm. In the case of grog, lower values of the amount of quartz were determined than in the raw illite. Quartz (or its high-temperature modifications) is a stable phase up to 1713 °C, but the melting temperature decreases in the presence of melting oxides such as K_2O which is contained in illite.

Table 5.4 – The amount of quartz in input materials as determined by the DSC method (in wt.%). The values in brackets are the amounts of quartz determined from the smaller peak on heat flow curve.

illite	grog	FFA	FBA	PFA
13 (0.86)	10 (0.70)	18	44	3.0

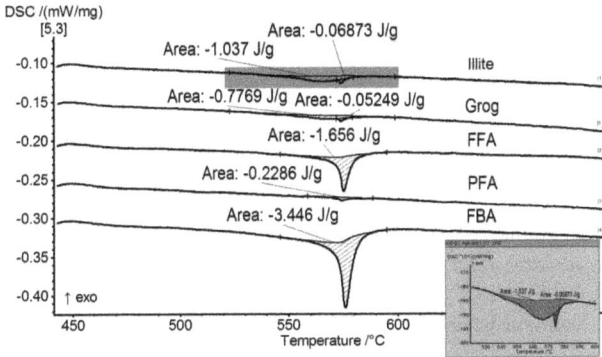

Fig. 5.8 – DSC analysis (heat flow curves) of input materials in the temperature area of
α-β modification transition of quartz. The inserted plot is a zoom on the endothermic
reaction of illite where a superposition of two peaks is visible. (the graphics is exported
from software Netzsch Protheus v.6.0.0).

5.1.4.2 Mineral composition of input materials

In Fig. 5.9 are the XRD spectra of the input materials[2]. The results of the DSC analysis
confirm that all materials contain a certain amount of quartz. Apart from illite and quartz,
no other minerals are observed in the illitic clay. However, it should be noted that the
differentiation of various phyllosilicates such as illite, muscovite, smectite,
montmorillonite, mica, and others can be problematic by XRD analysis due to the similar
structure and number of XRD reflections at nearby positions. After firing at 1100 °C and
90 minutes of isothermal holding (i.e. during the production of the grog), the illite
reflections completely disappeared. The quartz remained in the material, and about 11 wt.%
of the mullite was formed. The rest of the grog can be considered as an amorphous glassy
phase. In addition to quartz, FFA also contained a considerable amount of limestone, as
well as anorthite and gypsum. Most of the PFA is amorphous, which is well seen from the
weak reflections of the XRD spectrum. The crystalline component consists of anorthite,
quartz, and magnetite. From the first view of the XRD spectrum of FBA, it is clear that it
contains a considerable amount of quartz (44 wt.% according to DSC analysis). In addition
to it, anorthite, calcite, anhydrite are also present in this material. A weak reflection at an
angle of $2\theta = 18°$ corresponding to melanterite is also noticeable. A quantitative evaluation
of the proportion of individual minerals and the non-crystalline (or unidentified)
component is given in Table 5.5.

---

[2] Intensity unit a.u. is an abbreviation of the arbitrary unit. It is a procedurally determined unit used to compare the
results of measurements performed under the same conditions, most often in the same laboratory on the same
instrument.

Table 5.5 – Mineral composition of input materials (wt.%).

mineral	illite	grog	FFA	PFA	FBA
quartz	13	10	18	3.0	44
illite	66				
mullite		11			
calcite			11		0.51
gypsite			1.4		
anorthite			10	11	21
magnetite				4.4	
anhydrite					5.7
melanterite					1.3
amorf./undetermined	21	78	60	82	27

It is useful to know how the mineralogical composition of the raw materials will change if they are subjected to heat treatment. In Fig. 5.10 are the XRD spectra of illite, FFA, PFA, and FBA fired at 1100 °C at a rate of 5 °C min^{-1} and lasting 5 min at this temperature. It is therefore the same temperature program that was used to fire the samples of the investigated mixtures to the maximum temperature. On this basis, later in section 5.4, the behavior of the individual components individually and in mixtures will be analyzed. Grog was not fired because it is stable up to 1100 °C.

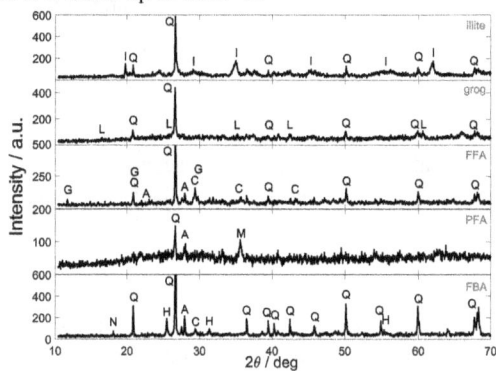

*Fig. 5.9 – XRD spectra of input materials for the preparation
of samples. The spectra of FBA and FFA are cropped for
better visualization. The intensity of the strongest reflection of
quartz in FBA is 2413 a.u. and of FFA 803 a.u. Q – quartz
(SiO$_2$), I – illite, L – mullite (3Al$_2$O$_3$·2SiO$_2$), G – gypsum
(CaSO$_4$·2H$_2$O), C – calcite (CaCO$_3$), A – anorthite
(CaAl$_2$Si$_2$O$_8$), M – magnetite (Fe$_3$O$_4$), H – anhydrite (CaSO$_4$),
N – melanterite (FeSO$_4$·7H$_2$O).*

The quantitative representation of minerals in the input raw materials after heating to a temperature of 1100 °C is summarized in Table 5.6. The amount of quartz in all materials remained unchanged. However, its percentage has changed due to the change in the weight of the material by heating. The body made of pure illite (without grog) contains 13 wt.% mullite and 12 wt.% sanidine (a high-temperature form of potassium feldspar) after firing at 1100 °C.

A significant amount of gehlenite (52 wt.%) was formed in FFA by heating to 1100 °C. It also contains anorthite (19 wt.%) and anhydrite (7 wt.%). Quantitative analysis shows that only a minimal amount of amorphous phase is present in the FFA after it is fired at 1100 °C (see Table 5.6).

Table 5.6 – Mineral composition of input materials after heating to 1100 °C (wt.%).

mineral	illite	FFA	PFA	FBA
quartz	14	22	3	48
mullite	13		24	
sanidine	12			
gehlenite		52		
anorthite(albite)		19	46	19
anhydrite		7		6
hematite			11	
amorf./undetermined	75	0	16	27

A large amount of crystalline phase was also formed in PFA. The magnetite originally contained in PFA is not observed after firing at 1100 °C. However, hematite (11 wt.%), mullite (24 wt.%), and anorthite/albite (46 wt.%) formed instead. In Fig. 5.10 there is another reflection (marked as "?") close to the anorthite peak, which has not been attributed to any mineral. Its position would suit the reflection of cristobalite, but it is unlikely that cristobalite would form at a temperature of 1100 °C.

In FBA, after firing at 1100 °C, the reflection on the calcite crystals disappeared (calcite decomposes at temperatures around 800 °C). The reflection that was attributed to melanterite also disappeared. The content of anhydrite and anorthite did not change by firing (only within the measurement uncertainty).

It must be acknowledged that quantitative evaluation of the proportion of minerals such as mullite, anorthite, gehlenite, sanidine, illite is associated with great uncertainty, due to the low intensity of their reflections on XRD spectra. More accurate values would require much more in-depth research. For these reasons, the relative uncertainty of the percentage of the above-mentioned minerals can reach an estimated 30%.

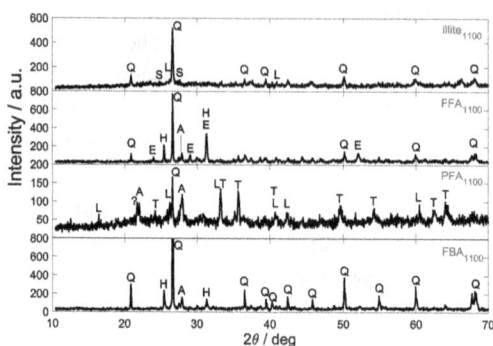

Fig. 5.10 – XRD spectra of input materiasl after heating to 1100 °C. Q – quartz (SiO_2), A – anorthite ($CaAl_2Si_2O_8$), H – anhydrite ($CaSO_4$), S – sanidine ($KAlSi_3O_8$), E – gehlenite ($Ca_2Al[AlSiO_7]$), T – hematite (Fe_2O_3), L – mullite ($2Al_2O_3 \cdot SiO_2$).

5.1.5   *Thermal analyses of input materials*

This subchapter presents the results of differential thermal analysis (DTA), thermogravimetry (TG), and evolved gas analysis (EGA) of input materials for sample preparation. Their analyses will provide an overview of the thermally induced processes and reactions. All three analyzes were performed simultaneously on powder samples with a mass of ~20 mg at a heating rate of 20 °C min^{-1} in a dynamic atmosphere of 21% O_2 + 79% Ar (see section 4.8 for further information).

5.1.5.1 *Illite*

In Fig. 5.11 is the DTA, TG, and DTG (time derivative TG) of powder illite. The same material including its preparation was used for the preparation of experimental samples. At the beginning of the heating, up to a temperature of ~200 °C, a weight loss of about 2% occurred. The endothermic minimum on the DTA curve correlates with this process. Considering the given temperature range, it can be argued that this is a release of physically bound water. This is also confirmed by the results of the EGA in Fig. 5.13, where the increased release of H_2O molecules is observed. As visible, there are some artifacts[3] on thermograms. The main flaw is the unexplained apparent increase in sample weight at some points of the TG curve, most probably due to the sample holder vibrating. DTA and TG measurements on compact samples, Fig. 5.12, have been shown to give seemingly better results. The measurements were performed on a Derivatograph 1100° [89], which, however, allows measurements only up to 1030 °C. A closer look at the measurement results of the compact sample shows that the process of weight loss in the temperature

---

[3] Artifact – in the terminology of thermal analyzes it is the effect observed on the measured curve (thermogram), which is not directly related to the processes in the sample [14]

range up to 250 °C has two minima on the DTG, DTA curves. Two peaks in this temperature range are also observed on the H_2O ion current curve of the EGA analysis in Fig. 5.13. The release of water molecules from the interlayer space of illite crystals seems to be a probable explanation. However, it can also be the dehydroxylation of gibbsite, which dehydroxylates at (210 – 240) °C [32]. It could be found in the material as a separate mineral. However, given the fact that the octahedral layer of illite is structurally identical to gibbsite (in fact, the octahedral layer is called "gibbsite layer"), it could also be a dehydroxylation of these layers of illite crystals. The fact that these layers, which are normally surrounded by tetrahedral layers in illite crystals at such low temperatures, can be explained by the fact that part of the gibbsite layers of illite is exposed, i.e. it is not surrounded by tetrahedral layers (for example if particles are small or defective). Verification of this hypothesis would require further research, but for this work and the production of illite-based ceramic materials, this phenomenon is not essential.

Immediately after the processes of releasing physically bound water, the process of burning (thermo-oxidation) of the organic material contained in the illite begins (~200 – 500 °C). This is best seen in the release of CO_2 molecules in Fig. 5.13. On the TG curve, this process manifests itself as a continuing weight loss, which at 500 °C reaches ~4.5%. From a temperature of about 420 °C, structurally bound water begins to be released from the illite during dehydroxylation. In it, OH-groups merge and release from the octahedral layer of the illite by 1D transport through the tetrahedral layer and subsequent 2D migration through the interlayer space [41, 97, 98]. The two observed minima of mass loss are due to the fact that the material is a combination of cis- and trans-vacant TOT layers of illite [98]. When heated to temperatures above 1000 °C, the ionic current (in EGA) from SO_2 molecules begins to increase. However, mass loss cannot be clearly confirmed based on this TG analysis. Although special minima and maxima are observed on the DTG curve at about 1100 °C, these may be possible artifacts. In addition, according to the results of the XRF analysis (see Table 5.2), the sulfur content of the illite is very small.

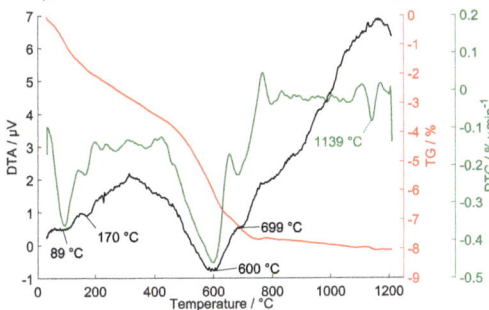

Fig. 5.11 – DTA (black), TG (red) and DTG (green) of illite. Powder sample. Atmosphere: 21 % O_2 + 79 % Ar, flow rate 60 ml min^{-1}. Heating rate: 20 °C min^{-1}.

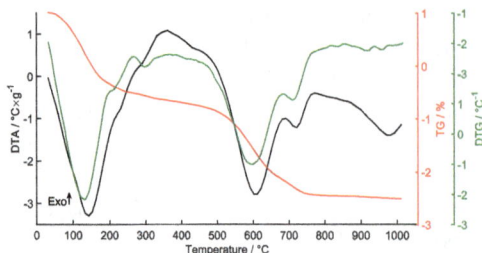

Fig. 5.12 – DTA (black), TG (red) and DTG (green) of illite. Compact sample ~3 g, static (self-generated) atmosphere. Apparatus: Derivatograph 1000°. Heating rate: 5 °C min^{-1}.

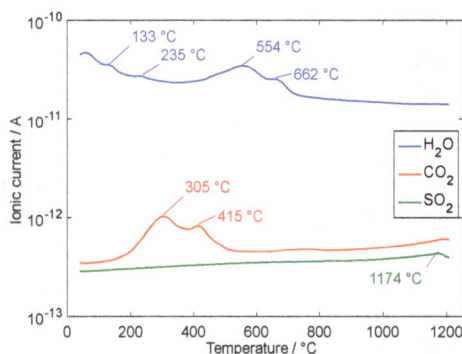

Fig. 5.13 – EGA of illite. Powder sample. Atmosphere: 21 % O_2 + 79 % Ar, flow rate: 60 ml min^{-1}. Heating rate: 20 °C min^{-1}.

5.1.5.2 FFA

In Fig. 5.14 is a graph of DTA, TG, and DTG of FFA used for the preparation of samples. The analysis of the released gases is shown in Fig. 5.15. As can be seen from the weight loss at low temperatures, FFA is able to bind a significant amount of moisture – about 9 wt.%. In addition, it is possible that some of the water molecules released in this temperature range come from the dehydration of gypsum ($CaSO_4 \cdot 2H_2O$) [99, 100]. The weight loss continued with increasing temperature even after these processes. Taking into account the results of the analysis of the released gases, it can be said that the given decrease is caused by the release of CO_2 molecules. Their origin is in the gradual heating of the rest of the organic material, especially the remnants of brown coal. In addition, an endothermic reaction associated with mass loss and the release of H_2O molecules is observed at ~400 °C. The most probable cause is the decomposition of calcium hydroxide

$(Ca(OH)_2)$ [35], which, however, does not occur in the FFA as mineral portlandite because it was not observed by XRD analysis (see section 0). The mass loss in the temperature range 200 °C – 600 °C is about 4%. Another significant process with a maximum at 782 °C is the decomposition of calcium carbonate $(CaCO_3)$ [36], which is apparently present mainly in the form of calcite. The weight loss is 3.54%, from which it can be concluded that FFA contains 7.9 wt.% $CaCO_3$ (based on the decomposition equation $CaCO_3 \rightarrow CaO + CO_2$ and atomic weights). Here we see a discrepancy with the result of XRD analysis, according to which there is up to 11 wt.% of calcite in FFA. However, due to the limited accuracy of the quantitative results of the XRD analysis, it can be said that this difference is within the measurement uncertainty. A weak exothermic reaction correlated

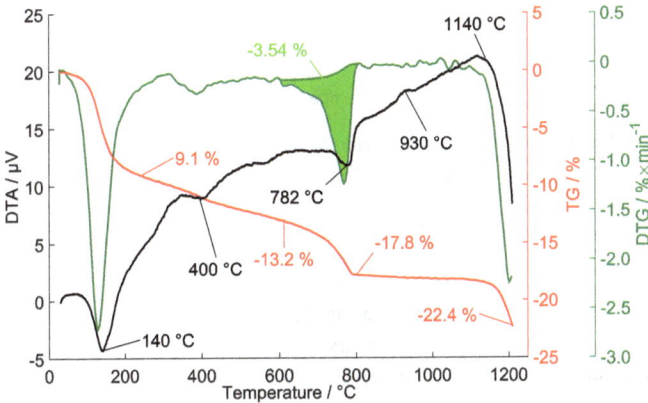

Fig. 5.14 – DTA (black), TG (red) a DTG (green) of FFA. Powder sample. Atmosphere: 21% O_2 + 79% Ar, flow rate: 60 ml min^{-1}. Heating rate: 20 °C min^{-1}

with CO_2 release was seen around 930 °C. This process can be associated with the recovery of residues of organic material, or carbon itself. However, it can also be a crystallization of gehlenite $(SiO_2 \cdot Al_2O_3 \cdot 2CaO)$, which is formed in FFA at high temperatures (see XRD Fig. 5.10). When the temperature rises above ~1140 °C, sulfur dioxide SO_2 begins to be released from the ash, which probably comes from the decomposition of calcium sulfate $CaSO_4$. Since SO_2 is potentially toxic, the temperature of 1140 °C can be considered as the maximum possible usable temperature for firing ceramic material with the addition of FFA.

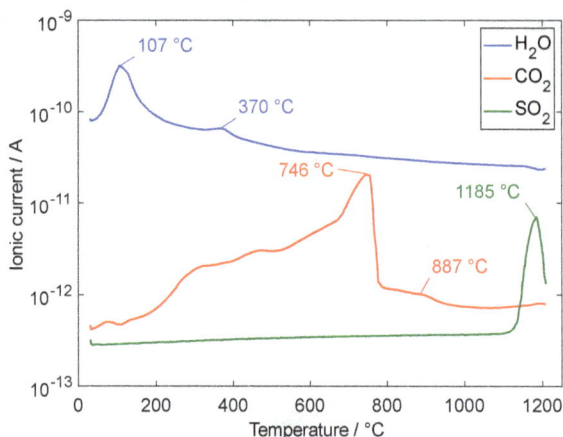

Fig. 5.15 – EGA of FFA. Powder sample. Atmosphere: 21% O_2 + 79% Ar, flow rate: 60 ml min^{-1}. Heating rate: 20 °C min^{-1}.

5.5.1.3 PFA

The total weight loss of PFA fly ash (see Fig. 5.16) up to 1200 °C was 4%. The mass loss was linear down to ~500 °C, so a typical drying interval cannot be recognized. Compared to FFA, PFA contained much less adsorbed air humidity. This result is consistent with its small specific surface area (3.7 g m^{-2}). The most significant mass loss associated with the exothermic reaction on the DTA curve is in the temperature range ~(400 – 600) °C. This is a thermo-oxidation of unburned lignite residues, as confirmed by the results of the analysis of released gases (see Fig. 5.17), where the release of CO_2 was observed. Another significant mass loss, which is again associated with the release of CO_2, had a maximum at 995 °C. The origin of this process is not entirely clear. The temperature is too high for the decomposition of $CaCO_3$ or the burning of organic components. From 1000 °C, the ionic current from SO_2 molecules increased slightly, but to a much lesser extent than in the case of FFA fly ash.

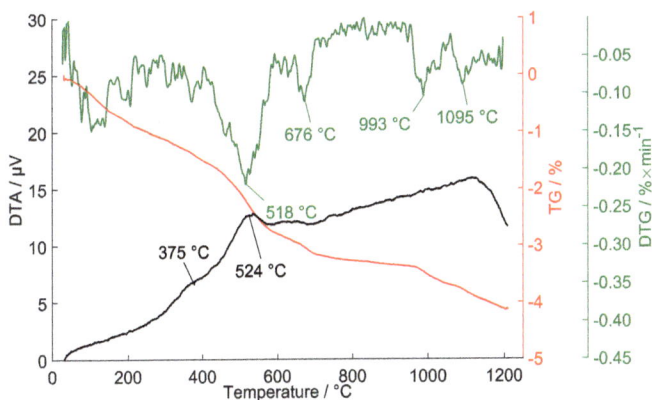

Fig. 5.16 – DTA (black), TG (red) a DTG (green) of PFA. Powder sample. Atmosphere: 21% O₂ + 79% Ar, flow rate: 60 ml min⁻¹. Heating rate: 20 °C min⁻¹.

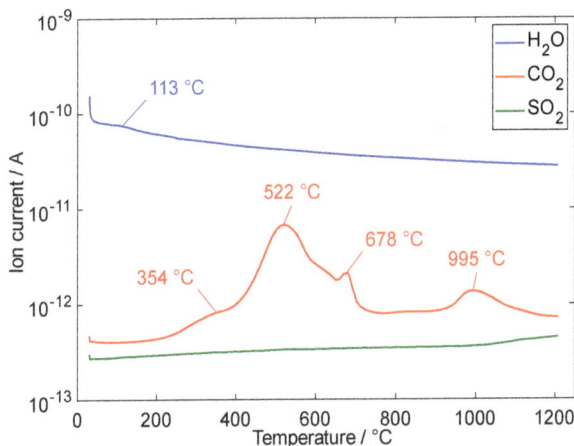

Fig. 5.17 – EGA of PFA. Powder sample. Atmosphere: 21% O₂ + 79% Ar, flow rate: 60 ml min⁻¹. Heating rate: 20 °C min⁻¹.

5.1.5.4 FBA

The moisture content in the bed ash from the fluidized bed boiler is 1.7 wt.% (see Fig. 5.18). At 463 °C, the mass loss occurred with the simultaneous release of water molecules (see Fig. 5.19). Accordingly, this process is a decomposition of $Ca(OH)_2$ [35] (however, portlandite was not observed on the XRD, so it is a non-crystalline form). The most significant is the mass loss due to $CaCO_3$ decomposition associated

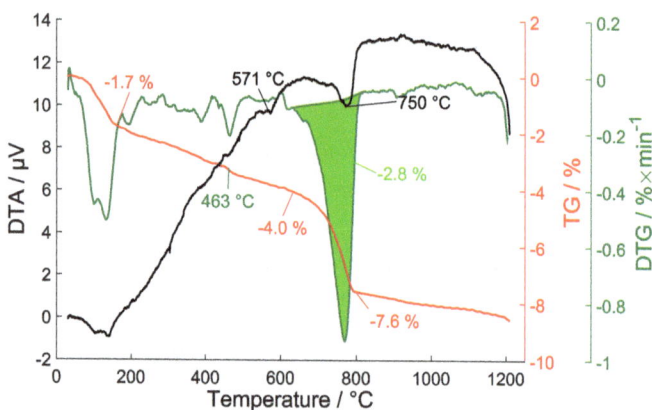

Fig. 5.18 – DTA (black), TG (red) a DTG (green) of FBA. Powder sample. Atmosphere: 21% O_2 + 79% Ar, flow rate: 60 ml min⁻¹. Heating rate: 20 °C min⁻¹.

with an endothermic reaction with a maximum at 750 °C. This reduced the weight by 2.8% (calculated by integrating the DTG peak), indicating that FBA contained 6.2 wt.% $CaCO_3$ (partly in the form of the mineral calcite). At 571 °C, there is an endothermic peak on the DTA curve, which is related to the α-β modification transition of quartz. In the temperature range between 150 °C and 650 °C, CO_2 is released, which comes mainly from the thermal oxidation of organic residues in the FBA. According to the EGA results (see Fig. 5.19) it can be seen that this process is relatively complex with several maxima. The total mass loss in this interval is 2.3%. At temperatures above 1100 °C, the ionic current from SO_2 molecules increased slightly but is smaller than in the case of FFA fly ash.

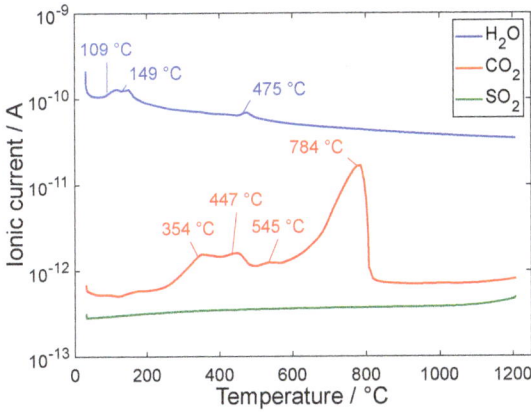

Fig. 5.19 – EGA of FBA. Powder sample. Atmosphere: 21% O_2 + 79% Ar, flow
rate: 60 ml min^{-1}. Heating rate: 20 °C min^{-1}.

5.2   Drying of wet ceramic bodies after shaping

The drying process was monitored on all experimental mixtures. Only the Bigot curve of the ceramic body made of pure illite is shown (see Fig. 5.20). The initial moisture (M_z) of the wet plastic mass for the preparation of samples was $(28.8 \pm 0.1)\%$. The slope (s) of the Bigot curve in the supercritical region is -0.54. This means that with each percentage of moisture, the sample shrinks by 0.54%. Drying can be considered safe after reaching a critical humidity $M_c = 11.24\%$. The drying sensitivity according to Bigot (BDS) of pure illite without impurities is 1.56 ± 0.01, so that the wet plastic mass of illite can be characterized as moderately sensitive to drying.

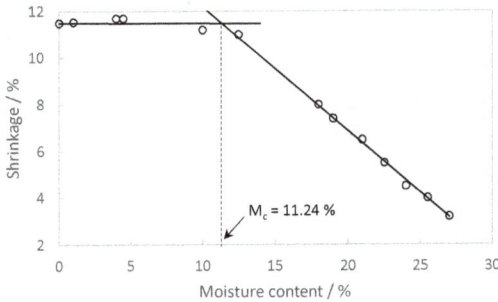

Fig. 5.20 – Bigod curve of ceramic body made from illite.

Table 5.7 – The values of initial humidity (M_z), critical humidity (M_c), supercritical slope (s), and Bigot drying sensitivity (BDS) for experimental mixtures. Values in parentheses are standard deviations at the position of the last digits (calculated from three values).

	Illite	A0		
M_z / %	28.8(1)	28.3(2)		
M_c / %	11.24(50)	16(3)		
s / %/%	−0.54(4)	−0.67(2)		
BDS / -	1.56(1)	0.78(28)		
	FF10	FF20	FF30	FF40
M_z / %	30.1(1)	33.4(3)	38.5(4)	37.5(2)
M_c / %	18.26(2)	19.83(95)	27.23(55)	24.63(63)
s / %/%	−0.72(8)	−0.66(4)	−0.76(4)	−0.66(1)
BDS / -	0.65(1)	0.69(9)	0.41(2)	0.52(4)
	PF10	PF20	PF30	PF40
M_z / %	31.2(1)	32.5(4)	33.3(3)	33.0(1)
M_c / %	16.74(34)	20.10(39)	19.97(81)	21.16(24)
s / %/%	−0.77(2)	−0.73(3)	−0.78(3)	−0.70(3)
BDS / -	0.87(5)	0.62(1)	0.67(6)	0.56(2)
	FB10	FB20	FB30	FB40
M_z / %	31.1(1)	31.1(1)	29.0(5)	31.1(1)
M_c / %	18.02(41)	18.33(83)	17.8(12)	19.62(51)
s / %/%	−0.82(1)	−0.79(5)	−0.68(4)	−0.78(5)
BDS / -	0.72(4)	0.70(8)	0.63(8)	0.59(4)

5.3　Differential thermal analysis

At present, DTA is performed almost exclusively on small amounts of powder samples (typically ~20 mg). This mass reduction leads to a decrease in temperature gradients in the volume of the examined sample. However, it could not be excluded that the reactions in compact samples differ from the reactions in powder samples due to the closer contact of the particles forming the sample. In order to obtain a more relevant conception of the processes in compact samples of the investigated mixtures, a Derivatograph 1100° instrument was used, on which it is possible to perform DTA of compact samples of relatively large dimensions (in this research, these were cylinders weighing about 3 g).

5.3.1　FFA mixtures

DTA results of mixtures of illite, grog, and FFA according to Table 4.4 are in Fig. 5.21. Fig. 5.22 represents the results of the thermal derivation of thermogravimetry - DTG analysis which better illustrates the processes related to the mass change. Reference sample A0 contains only illite and grog. Grog is stable in the temperature range of these measurements (up to 1030 °C), and therefore the DTA results are qualitatively identical to the results for pure illite (see Fig. 5.12). The only difference is in the size and area of the individual peaks (0.6 times the DTA peaks of pure illite). In all examined mixtures, an endothermic reaction is observed at the beginning of the heating process, which intensifies with an increasing amount of FFA. This reaction is related to the release of physically bound water (moisture) and correlates with DTG analysis. It can also be said that the FFA has a moisture release temperature shifted to higher values (169 °C vs. 121 °C) compared to illite. This shift can be explained, especially in the case of compact bodies, by the increased vapor pressure in the pores of the material and the furnace space (due to the static atmosphere used in the measurements). Endothermic reactions at 206 °C and 281 °C are due to the presence of illite and have been explained in section 5.1.5.1. Near 310 °C, an apparent exotherm is visible for FF40 and FF10 samples. However, since no correlation is observed between this exothermic peak and the FFA content, its origin cannot be explained yet. At a temperature of about 434 °C, the presence of FFA is manifested by the decomposition of the $Ca(OH)_2$ contained in it (see paragraph 5.1.5.2). As the temperature increases further, the illite dehydroxylates, resulting in an endothermic reaction and an acceleration of weight loss. Around 780 °C, the decomposition of $CaCO_3$ begins, manifesting itself in an endothermic reaction associated with weight loss. Again, a shift of the peak maximum on the DTA and DTG curves to higher temperatures is observed with increasing FFA content. This can be explained by the increase in CO_2 pressure generated by the decomposition of $CaCO_3$. All the processes observed so far have been more or less just a superposition of the input processes. However, at 920 °C, a significant exothermic reaction occurs, which does not correlate with any weight loss or increase. It is a process of crystallization of anorthite ($CaAl_2Si_2O_8$), which was formed from metaillite and FFA components [16, 101] (see XRD results in section 5.4). This process is probably due to the suitable stoichiometric ratio of the elements Si, Al, and Ca at the interface of the particles since the temperature is still too low for intensive crystallization from the melt to take place. Anorthite is welcome in ceramic materials because it increases strength and reduces firing shortening [102]. Furthermore, a slight decrease in the DTA curve can be seen at temperatures approaching 1000 °C, which can be attributed to the melt formation.

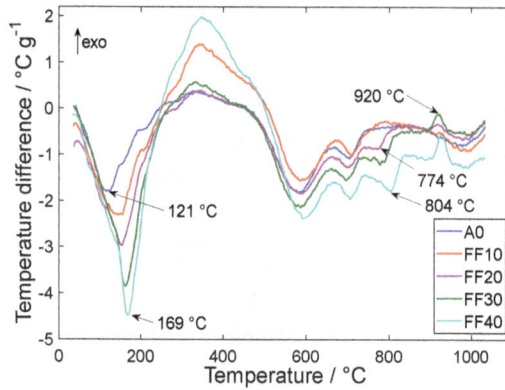

Fig. 5.21 – DTA analysis of mixtures with 10 – 40 wt.% of FFA and a reference sample A0. Parameters: compact sample with a mass of ~3g; heating rate of 5°C min^{-1}; static air atmosphere.

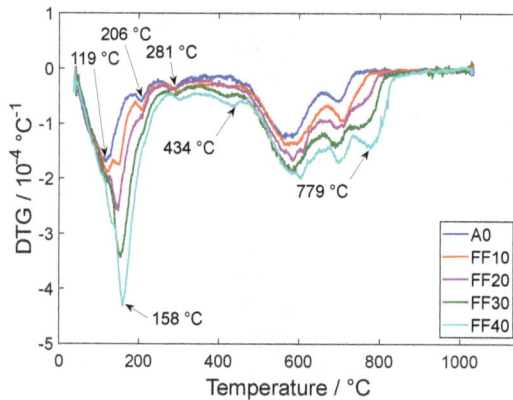

Fig. 5.22 – DTG analysis of mixtures with 10 – 40 wt.% of FFA and a reference sample A0. Parameters: compact sample with a mass of ~3g; heating rate of 5°C min^{-1}; static air atmosphere.

5.3.2   PFA mixtures

The behavior of drying of bodies with the addition of PFA, in contrast to mixtures with FFA, is given almost exclusively by drying of illite (see Fig. 5.23 and Fig. 5.24). This is due to the small amount of physically bound water on the surface of the PFA fly-ash particles (due to the small specific surface area of 3.7 m^2 g). At a temperature region of ~420 – 500 °C, an exothermic reaction occurs, the intensity of which is directly proportional to the amount of PFA. Although no clear correlation is observed on DTG with weight loss, the results of the analysis of the released gases (see Fig. 5.17) with respect to the amount of CO_2 released indicate that this is a thermo-oxidation of organic parts, most likely unburned brown coal residues. In addition to this process, the DTA and DTG curves for all mixtures with PFA are practically identical, and thus in the temperature range up to 1030 °C, PFA behaves similarly to the grog.

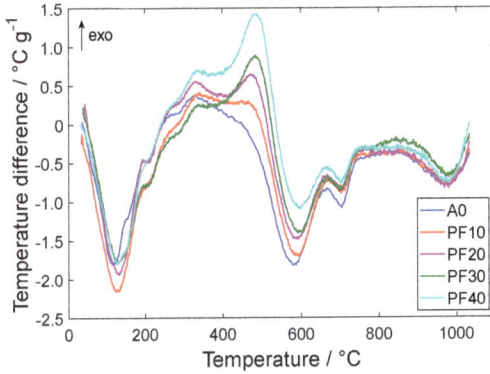

Fig. 5.23 – DTA analysis of mixtures with (10 – 40) wt.% of PFA and a reference sample A0. Parameters: compact sample with a mass of ~3g; heating rate of 5°C min⁻¹; static air atmosphere.

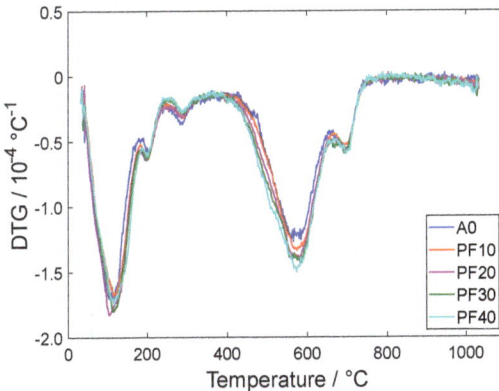

Fig. 5.24 – DTG analysis of mixtures with 10 – 40 wt.% of PFA and a reference sample A0. Parameters: compact sample with a mass of ~3g; heating rate of 5°C min⁻¹; static air atmosphere.

5.3.3 *FBA mixtures*

FBA fly-ash, like FFA, slightly shifts the maximum drying temperature to higher temperatures while slightly increasing the intensity of the endothermic reaction and the amount of moisture released (see Figs. 5.25 and 5.26). The decomposition of $Ca(OH)_2$ at ~460 °C (see Fig. 5.18) has little effect on DTA. However, the decomposition process of $CaCO_3$ is significant, of which 6.2% by weight is present in FBA.

The general conclusion regarding the results of DTA and DTG analyses of the investigated mixtures in the temperature range up to 1030 °C is that they can be considered a superposition of these analyses for individual starting materials. The only exception is the crystallization of anorthite in mixtures containing illite and FFA.

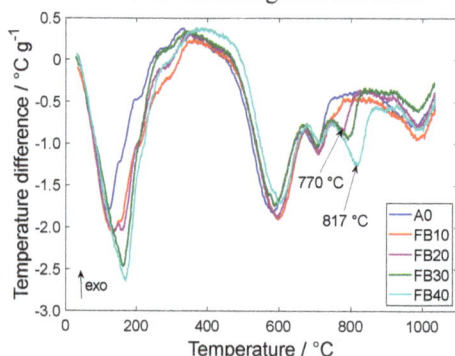

Fig. 5.25 – DTA analysis of mixtures with 10 – 40 wt.% of FBA and a reference sample A0. Parameters: compact sample with a mass of ~3g; heating rate of 5°C min⁻¹; static air atmosphere.

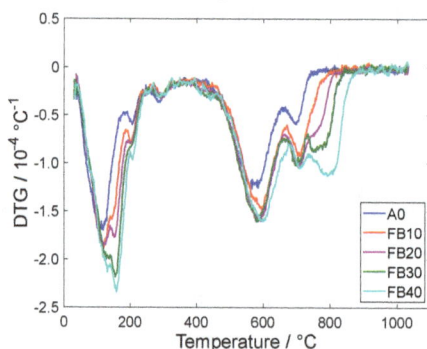

Fig. 5.26 – DTG analysis of mixtures with 10 – 40 wt.% of FBA and a reference sample A0. Parameters: compact sample with a mass of ~3g; heating rate of 5°C min⁻¹; static air atmosphere.

5.4 XRD analysis

The mineralogical composition of the individual mixtures directly after shaping and drying is a superposition of the mineralogical composition of the raw materials. In order for new mineral shapes to form, the heat supplied to the material must be sufficient to decompose the original minerals and create new ones. The formation of new minerals can be expected first above the warm onset of solid-phase sintering, i.e., more than ~800 °C. Due to limited access to XRD analysis, the study of mineralogical composition was limited to experimental mixtures fired at 1100 °C. At this temperature (the highest which was used to burn the samples), a detectable amount of new minerals is most likely to occur in the material. The firing regime used was the same as in the other analyzes: heating at a rate of 5 °C min^{-1} to a temperature of 1100 °C → 5 min isothermal heating → cooling to room temperature at a rate of 10 °C min^{-1}. Compact bodies were fired and pulverized for XRD analysis. In order to see whether new minerals were formed or whether the mineralogical composition of the mixtures fired at 1100 °C was only a superposition of the mineralogical composition of the precursors fired at 1100 °C, XRD spectra would also be calculated by mixing 60 wt.% of calcined illite and 40 wt.% of calcined ash. They were calculated as the sum of 0.6 times the XRD intensity of the calcined illite and 0.4 times the intensity of the calcined ash[4].

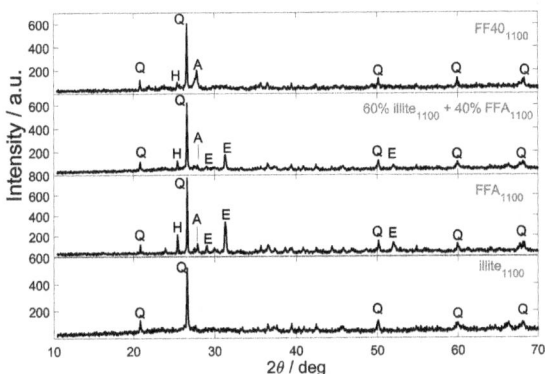

Fig. 5.27 – XRD spectra of a mixture of FF40 and its initial raw materials after firing at 1100 °C. The calculated XRD spectrum of a mixture of 60 wt.% of calcined illite and 40 wt.% of calcined FFA is also shown. Q - quartz (SiO_2), A - anorthite ($CaAl_2Si_2O_8$), H - anhydrite ($CaSO_4$), E - gehlenite ($Ca_2Al[AlSiO_7]$).

---

[4] Calculation of the mixed XRD spectrum in this way is possible because identical experimental conditions were used to record all XRD spectra. The correction for firing weight loss was not made because a simplified calculation is sufficient for this purpose.

In Fig. 5.27 the XRD spectrum of the FF40 mixture after firing at 1100 °C is depicted. It contains 17 wt.% of quartz, 67 wt.% of anorthite, and 4 wt.% of anhydrite. Other possible substances are corundum, sanidine, and lime. From the results of XRD analysis, it is clear, that the mineralogical composition of the FF40 mixture is not a simple superposition of the mineralogical composition of the initial materials. If the FFA was fired independently, a large amount of gehlenite (52 wt.%) was treated in it. It also contains anhydrite (7 wt.%) and anorthite (19 wt.%). If such a fired mixture is FF40, the formation of gehlenite is completely blocked, the anhydrite changes only to a small amount and, conversely, the formation is promoted. The overall chemical composition of the FF40 mixture does not completely correspond to the chemical composition of the anorthite, so it can be assumed that the anorthite is formed only in other areas, apparently at the range of illusion particles and calcite or anhydrite.

As can be seen from the XRD spectra in Fig. 5.28, the mineralogical composition of the PF40 mixture after firing at 1100 °C is essentially a superposition of the mineralogical composition of the precursors fired at 1100 °C. No new mineral phases were formed. The only change in the case of firing a mixture rather than individual components is the weakening of the reflections attributable to the anorthite. These are essentially lost in the experimental noise, making a quantitative evaluation of the anorthite contribution impossible. After firing to 1100 °C, the PF40 mixture contains 9.6 % quartz, 9.3 % mullite, 3 % hematite, and an unknown proportion of anorthite.

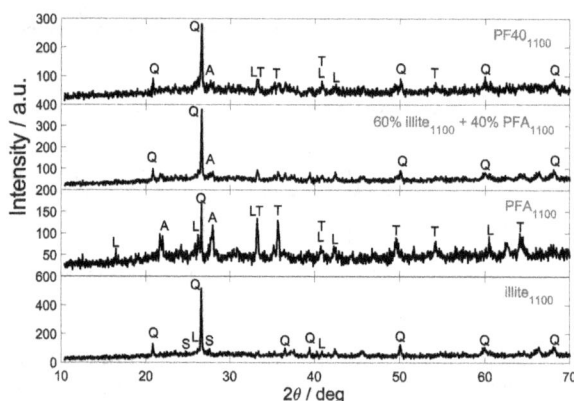

Fig. 5.28 – XRD spectra of a mixture of PF40 and its initial raw materials after firing at 1100 °C. The calculated XRD spectrum of a mixture of 60 wt.% calcined illite and 40 wt.% calcined PFA is also shown. Q - quartz (SiO_2), A - anorthite ($CaAl_2Si_2O_8$), T - hematite (Fe_2O_3), L - mullite ($2Al_2O_3SiO_2$), S - sanidin ($KAlSi_3O_8$).

Materials Research Forum LLC
https://doi.org/10.21741/9781644902073

The XRD spectra of the FB40 mixture after firing at 1100 °C as well as the XRD spectra of the feedstocks for the production of the FB40 mixture after firing at 1100 °C are shown in Fig. 5.29. By comparing the calculated spectra and the spectra of the fired FB40 mixture, it can be seen that the mineralogical composition of the FB40 mixture after firing at 1100 °C is a simple superposition of the mineralogical composition of the feedstocks. The same result can be expected for mixtures with less FBA fly-ash content. Quantitative analysis revealed that after firing at 1100 °C, the FB40 mixture consists of 27 % quartz, 22 % anorthite, 4 % anhydrite, and the remaining 47 % is attributed to the amorphous phase. The amount of mullite and sanidine could not be determined.

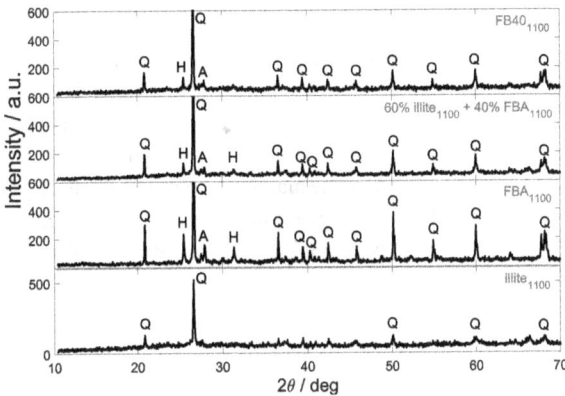

Fig. 5.29 – XRD spectra of the FB40 mixture and its feedstock after firing at 1100 °C. The calculated XRD spectrum of a mixture of 60% fired illite, and 40% fired FBA is also shown. The spectra are cropped for better display.
Q - quartz (SiO$_2$), A - anortite (CaAl$_2$Si$_2$O$_8$), H - anhydrite (CaSO$_4$).

XRD analysis showed that after firing at 1100 °C, the mineralogical composition of the mixtures of illite and PFA and FBA fly-ashes is the sum of the mineralogical compositions of the individual components of the mixtures. The coupling of these components by the wet-forming process did not allow sufficiently close contact of the individual phases to affect the formation of the novel minerals. This suggests that the mineralogical changes occur within the individual particles (grains, agglomerates) of the individual components of the investigated mixtures. A different situation was observed in the mixtures of illite and FFA fly-ash. Their combination prevented the formation of the mineral gehlenite in the FFA and promoted the formation of anorthite in the FF40 mixture. Since significant melt-formation is not observed (see Fig. 5.97), the formation of gehlenite and anorthite appears to occur predominantly in the contact zones of the particles containing CaO (originally

$CaCO_3$) and the Al_2O_3 and SiO_2 oxides. A similar phenomenon has not been observed for PFA fly-ash. The reason is that it contains only a small amount of CaO. In the case of FBA, the reason is probably inappropriate stoichiometry in the contact zones or insufficiently contact between illite and FBA fly-ash particles.

5.5 Mass changes

5.5.1 FFA mixtures

The relative mass change during heating (thermogravimetry, TG) for samples containing different amounts of FFA fly-ash is shown in Fig. 5.30. The results correspond with the DTG results in Fig. 5.22. FFA fly-ash has the ability to absorb large amounts of moisture (see Fig. 5.14), which is also evident in its mixtures with illite. Another significant decrease in mass is in the dehydroxylation region of illite. The relative weight changes for the same mixtures after firing to different temperatures are shown in Fig. 5.31. The results are similar to those obtained during heating. The small differences can be attributed to the fact that a 5 min isotherm was always applied at the maximum temperature during firing. When firing the samples at 1100 °C, it can be seen from Fig. 5.31 that the mass loss from the decomposition of $CaSO_4$ is already beginning to show. For this reason, the maximum firing temperature should be 1000 °C.

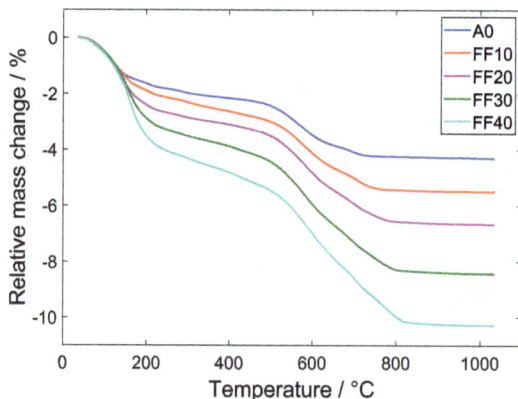

Fig. 5.30 – Relative mass change during heating of mixtures with 10 to 40 % FFA and a reference mixture A0.

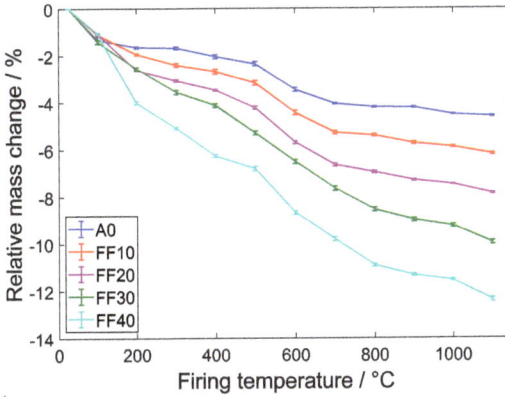

Fig. 5.31 – Relative mass change after firing to different temperatures of mixtures with 10 to 40 % FFA and a reference mixture A0.

5.5.2   PFA mixtures

The results in section 5.3 indicated, there are no significant changes in the PFA in the temperature range up to 1100 °C. This behavior is also evident in the results for the relative mass change of the mixture of PFA, grog, and illite in Fig. 5.32 and Fig. 5.33. The mass change of the samples containing PFA fly-ash is practically identical to the reference sample A0. Its relative mass changes are conditioned by the mass changes of illite because the grog is temperature stable. This behavior of the PFA fly-ash can be regarded as positive since, if the firing does not generate gases, there is no danger of rapid firing increasing the pressure in the body volume above the tolerable level.

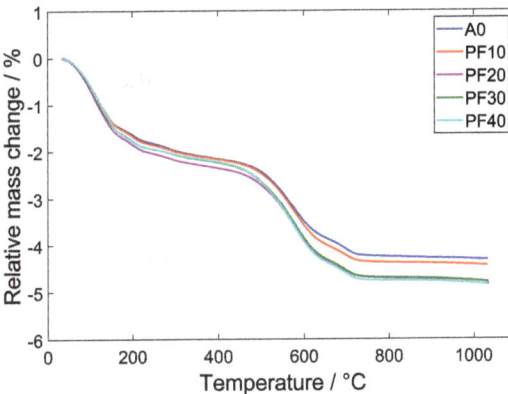

Fig. 5.32 – Relative mass change during heating of mixtures with 10 to 40 % PFA and a reference mixture A0.

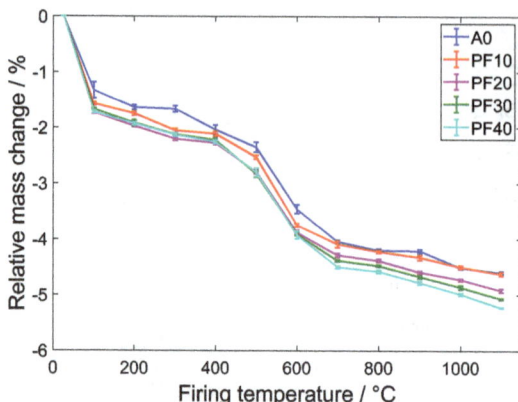

*Fig. 5.33 – Relative mass change after firing to different temperatures of mixtures with
10 to 40 % PFA and a reference mixture A0.*

5.5.3   *FBA mixtures*

Considering the results of the relative mass loss during heating (see Fig. 5.34) and after firing (see Fig. 5.35) of the mixtures containing FBA fly-ash, it can be said that the amount of moisture contained in the samples containing FBA fly-ash is greater than that of the samples containing PFA fly-ash but less than that of the mixtures containing FFA fly-ash (which is consistent with the moisture content of the individual fly-ashes). The mass change associated with moisture release measured after firing does not correspond to the mass change measured during heating. The difference is up to two percent, which can be explained by the different initial moisture contents of the samples used. As with the previous mixtures, the significant decrease in mass is associated with the dehydroxylation of illite in the temperature range ~500 – 700 °C. $CaCO_3$ decomposition also contributes to the weight loss of about 1 wt.% (in the case of FB40). Similar to the FFA fly-ash, a mass loss due to $CaSO_4$ decomposition is observed when the samples are fired at 1100 °C so that the maximum firing temperature must be lower than 1100 °C.

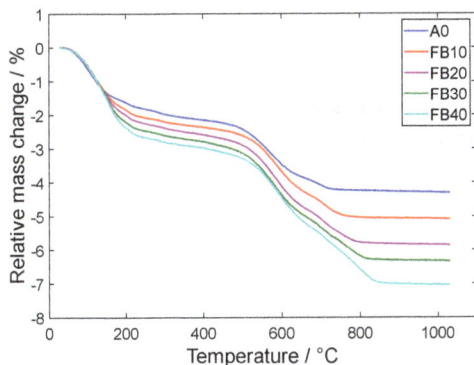

Fig. 5.34 – Relative mass change during of mixtures with 10 to 40 % FBA and a reference mixture A0.

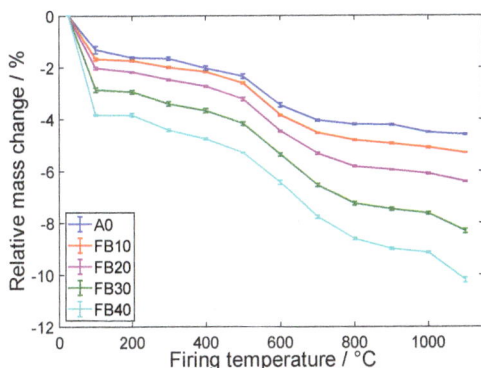

Fig. 5.35 – Relative mass change after firing to different temperatures (b) of mixtures with 10 to 40 % FBA and a reference mixture A0.

It can be verified by calculation that the mass changes of the individual mixtures are a simple superposition of the mass changes that are measured on the input raw materials. An important finding is that the maximum firing temperature of samples containing FFA and FBA fly-ash (i.e., fly-ash from fluidized bed boilers) must be lower than the $CaSO_4$ decomposition temperature ~1100 °C.

5.6   Dimension changes

5.6.1   *FFA mixtures*

The relative dimensional change at actual temperature during firing (thermodilatometry, TD) for reference sample A0 and mixtures containing FFA fly-ash is shown in Fig. 5.36. Up to a temperature of ~900 °C, the TD curves of all mixtures are essentially identical. The same can be said for the relative dimensional change measured at room temperature (see Fig. 5.37). In this temperature range, the most pronounced increase in dimensions of about 1 % is due to dehydroxylation of illite.

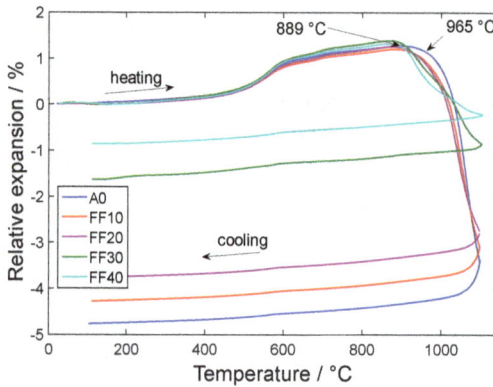

Fig. 5.36 – Relative length change during heating of mixtures with 10 to 40 % of FFA and a reference mixture A0.

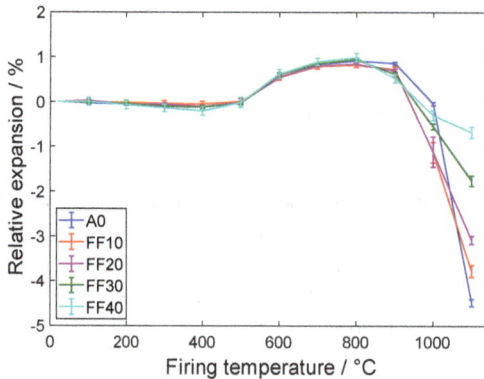

Fig. 5.37 – Relative length change after firing to different temperatures of mixtures with 10 to 40 % of FFA and a reference mixture A0.

The two-step nature of dehydroxylation is also reflected in the increase in dimensions, as seen from the derivative TD (DTD) curves in Fig. 5.38. The rapid sintering of sample A0 starts at 965 °C. The addition of FFA fly-ash lower this temperature to 889 °C for a 40 % fraction. According to the DTD curves (see Fig. 5.39) of mixtures FF30 and FF40, it can be said that the addition of FFA results in an additional sintering step that precedes that observed for the reference mixture A0. Its maximum is at 930 °C, which correlates with the formation of anorthite (see Fig. 5.14). The A0 mixture sinters most intensely (see Fig. 5.38), and the sintering intensity decreases with the addition of FFA. This is associated with a significant reduction in shrinkage by firing from 4.5% for the A0 sample to 0.7% for the FF40 sample (see Fig. 5.37). The reduction in sintering temperature, sintering intensity, and shortening by firing, can be considered as a positive effect of the replacement of the conventional temper by FFA fly-ash. During cooling, the DTD curves show a modification transformation of the quartz, which is associated with a reduction of its dimensions near the temperature of 580 °C (see Fig. 5.39). The dimensional change was more pronounced when the mixture contained 30 and 40 % of FFA. Such shrinkage of quartz crystals causes the formation of microcracks that are partially closed on further cooling if the material includes a sufficient amount of enamel [103]. The effect of illite dehydroxylation masked the modification transformation of quartz during heating.

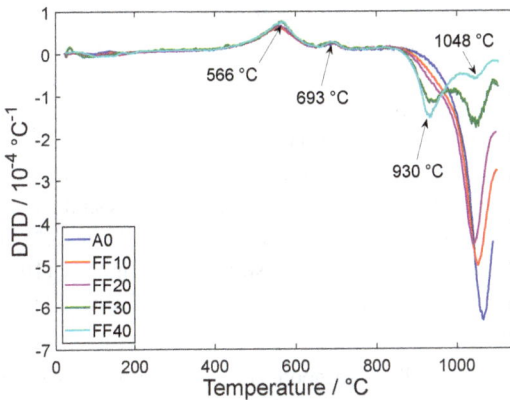

Fig. 5.38 – Thermal expansion coefficient (temperature derivative of TD) of mixtures with 10 to 40 % FFA and a reference mixture A0 during heating.

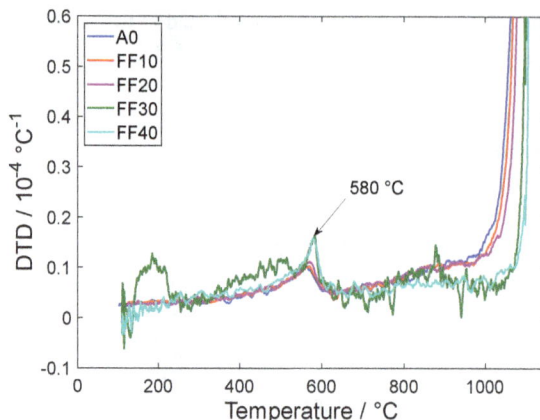

Fig. 5.39 – Thermal expansion coefficient (temperature derivative of TD) of mixtures with 10 to 40 % FFA and a reference mixture A0 during cooling.

5.6.2   PFA mixtures

Fig. 5.40 shows TD measurements of mixtures containing PFA fly ash. As with the addition of FFA, up to a temperature of 900 °C, the dimensional changes are almost exclusively due to the dimensional changes of illite. However, at temperatures above 900 °C, the effect of PFA fly-ash on the dimensional changes is substantially different from that of FFA. The addition of PFA increases the shrinkage by firing already after firing to 1000 °C, as can be seen from Fig. 5.41. This effect becomes even more pronounced after firing to 1100 °C, see Fig. 5.40 and Fig. 5.41. From the DTD curves in Fig. 5.42, it is clear that PFA increases the sintering intensity and shifts the inflection point systematically from 1066 °C for the reference mixture A0, to 1040 °C for the PF40 mixture. This shift can be explained by the presence of the melting oxides K_2O and Na_2O (see Table 5.2). By replacing the conventional PFA abrasive with fly-ash, the shrinkage by firing at 1100 °C increases from 4.5 % to 8.2 %. As suggested by the results, additional PFA would probably not increase the shrinkage any further.

PFA has very good sinterability, so good mechanical properties of the bodies with its addition can be assumed. The reduction of the sintering temperature by about 46 °C can also be considered as a positive property. The disadvantage is the increase in overall shrinkage by firing.

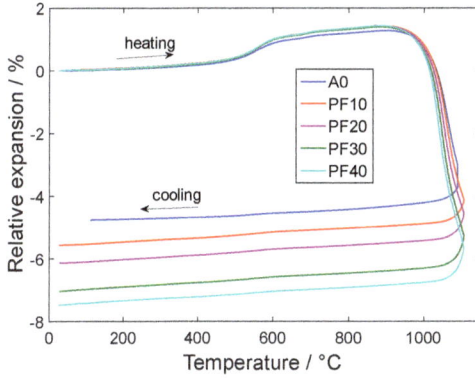

Fig. 5.40 – Relative length change during heating of mixtures with 10 to 40 % PFA and a reference mixture A0.

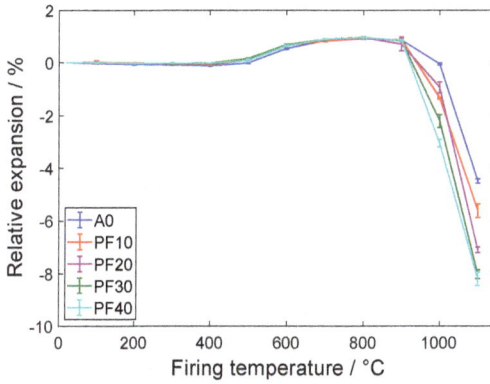

Fig. 5.41 – Relative length change after heating to different temperatures of mixtures with 10 to 40 % PFA and a reference mixture A0.

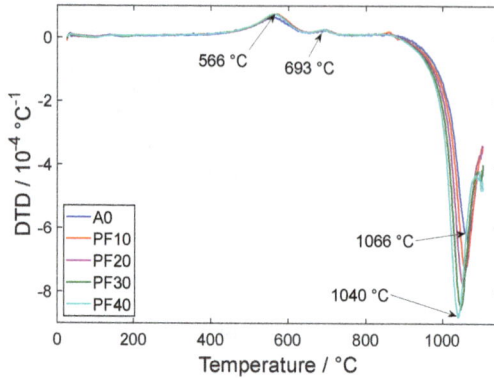

Fig. 5.42 – Thermal expansion coefficient (temperature derivative of TD) of mixtures with 10 to 40 % PFA and a reference mixture A0 during heating.

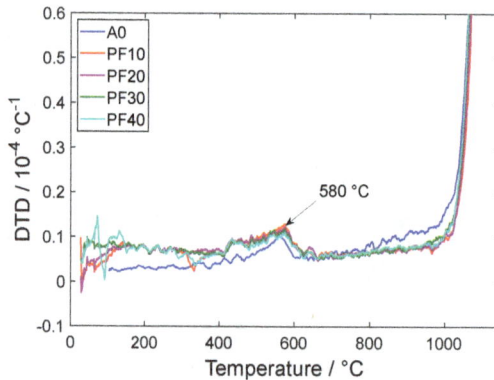

Fig. 5.43 – Thermal expansion coefficient (temperature derivative of TD) of mixtures with 10 to 40 % PFA and a reference mixture A0 during cooling.

5.6.3   FBA mixtures

The FBA fly-ash contains a significant amount of quartz (44%), observable during heating at around 570 °C as an additional increase in dimension (see Figs. 5.44 and 5.46), which is added to the increase in dimension caused by the dehydroxylation of illite. After firing the samples to 1100 °C, mixtures containing 10 and 20 wt.% FBA achieve more significant shrinkage than the reference mixture A0. However, when the mixture has 30, and 40 wt.% FBA, the shrinkage by firing is significantly reduced (see Fig. 5.45). The reason for such a special dependence on the proportion of FBA in the mixtures is not yet apparent. The discrepancy between the overall shrinkage after firing to 1100 °C observed in Fig. 5.45 and the TD results in Fig. 5.44 may be explained by the additional isothermal segment of the temperature program that was applied to the firing of the samples used to measure the dimensional change by firing at room temperature. The quartz present to a large extent in the FBA fly-ash also exhibits abrupt contraction on cooling (see Fig. 5.47). In the same figure, at temperatures below 300 °C, there are peaks that are not related to actual changes in the dimensions of the samples, but are experimental errors (artifacts) that have been amplified by numerical derivation of the TD curve.

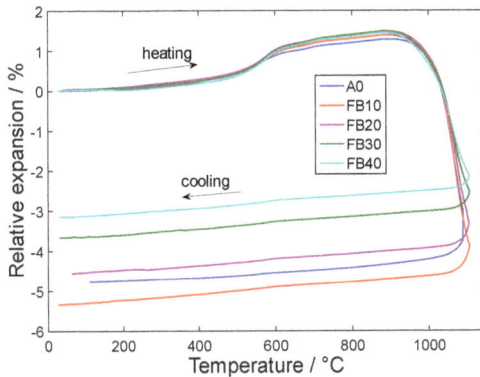

Fig. 5.44 – Relative length change during heating of mixtures with 10 to 40 % FBA and a reference mixture A0.

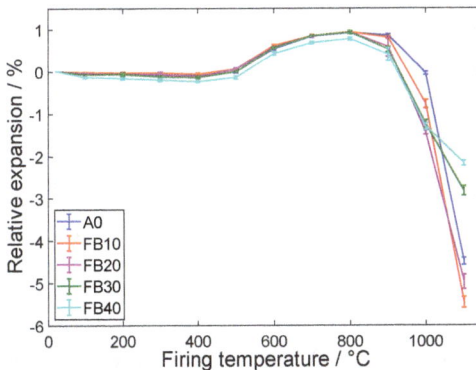

Fig. 5.45 – Relative length change after firing to different temperatures of mixtures with 10 to 40 % FBA and a reference mixture A0.

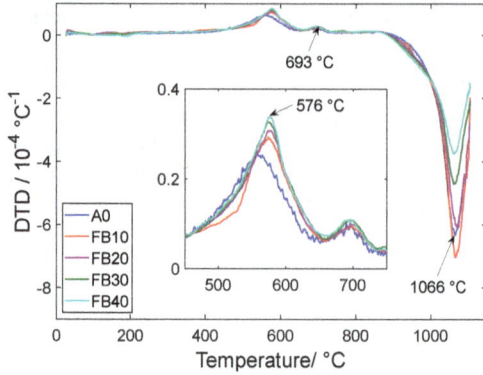

Fig. 5.46 – Thermal expansion coefficient (temperature derivative of TD) of mixtures with 10 to 40 % FBA and a reference mixture A0 during heating.

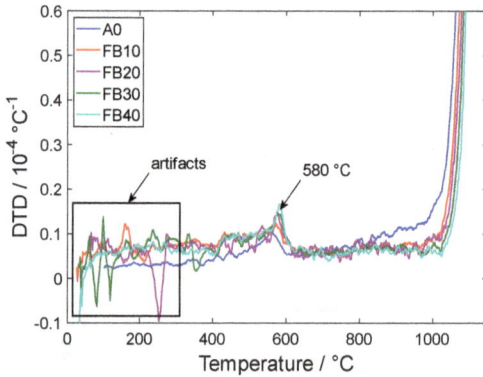

Fig. 5.47 – Thermal expansion coefficient (temperature derivative of TD) of mixtures with 10 to 40 % FBA and a reference mixture A0 during cooling.

5.7   Bulk density

5.7.1   *FFA mixtures*

The evolution of the bulk density ρ during heating of mixtures containing 10 to 40 wt.% FFA and the reference mixture A0 is shown in Fig. 5.48. The cooling is in Fig. 5.49. The bulk density measured at the actual temperature during heating and cooling was calculated based on the TD and TG measurements and the initial ρ. At first glance, it can be seen that the addition of FFA reduces ρ both before and after firing (see Fig. 5.50). Low ρ is considered a positive property in construction materials (while maintaining adequate mechanical strength). The difference between the raw samples A0 and FF40 is almost 0.3 g cm^{-3}, about 20%. Drying causes a reduction in ρ, and the more fly-ash the mixture contains, the more pronounced this reduction is. The dehydroxylation of illite and the decomposition of $CaCO_3$ have a similar effect. Anorthite formation and sintering, on the other hand, increase bulk density. During cooling, the bulk density does not change significantly; only the reduction of the dimensions of all the bodies by cooling is apparent. After firing, the FF40 mixture is ~0.55 g cm^{-3} smaller than A0, a difference of about 42%. The bulk density measured after firing is qualitatively similar to that measured during heating. The slight difference between them is due to the reversible thermal expansion of the material.

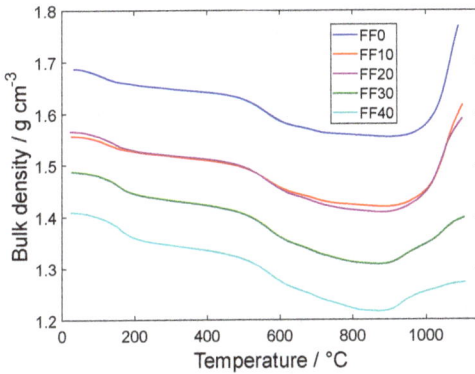

Fig. 5.48 – Bulk density of mixtures with 10 to 40 % FFA and reference mixture A0 during heating.

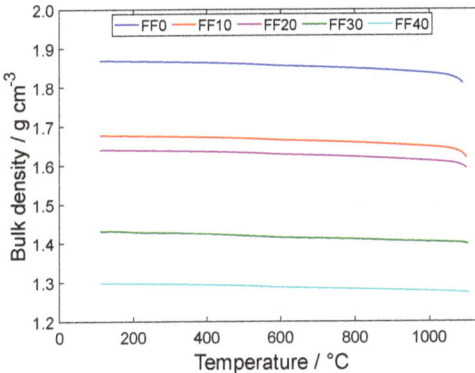

Fig. 5.49 – Bulk density of mixtures with 10 to 40 % FFA and reference mixture A0 during cooling.

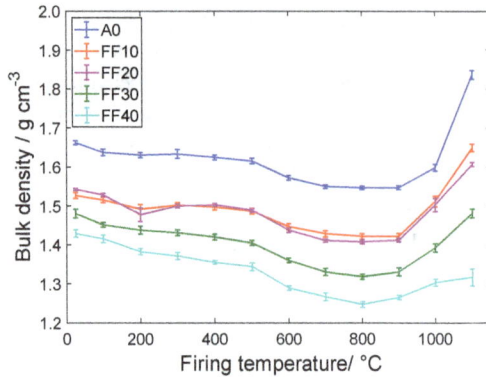

Fig. 5.50 – Bulk density of mixtures with 10 to 40 % FFA and reference mixture A0 after firing at different temperatures.

5.7.2   PFA mixtures

The evolution of the bulk density of the mixtures containing PFA fly-ash during heating and cooling is shown in Figs. 5.51 and 5.52. Similar to FFA, the addition of PFA to the mixtures reduces the bulk density of the green bodies. The exception is the PF10 mixture, which up to 900 ° C practically copied the development of the reference mixture A0. The explanation why the ρ of PF10 mixture is not lower than the ρ of A0 mixture from the start is not known. This phenomenon could possibly be due to different pressures during extrusion of the samples (as a consequence of the different plasticity of the mixtures during the extrusion), but so far, such an explanation is only a hypothesis. Drying and dehydroxylation of illite reduce the ρ approximately equally for all mixtures. During sintering, the ρ of PFA-containing mixtures begin to catch up with the ρ of the reference mixture A0 due to the more significant contraction of the PFA mixtures (see Fig. 5.40). This effect is even more pronounced when measured at laboratory temperature after firing (see Fig. 5.53). However, it should be kept in mind that when firing samples for measurements at laboratory temperature, a 5 min isothermal time was applied at the maximum temperature.

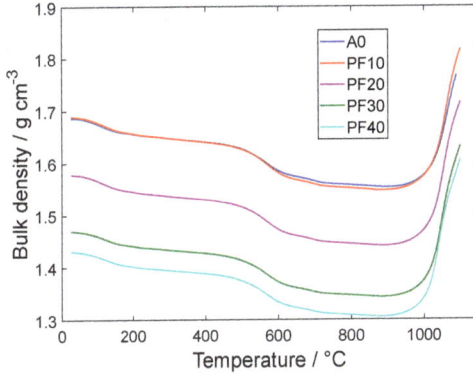

Fig. 5.51 – Bulk density of mixtures with 10 to 40 % PFA and reference mixture A0 during heating.

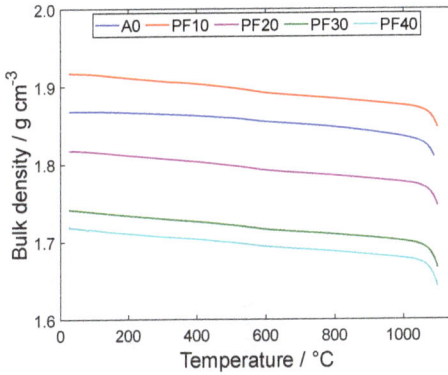

Fig. 5.52 – Bulk density of mixtures with 10 to 40 % PFA and reference mixture A0 during cooling.

5.7.3   FBA mixtures

The addition of 10 to 30 wt.% of FBA has practically no effect on the values of ρ up to 1000 °C (see Fig. 5.54). Qualitatively, the same trend of ρ was observed for the FB40 mixtures, but as can be seen, its values are shifted to lower values by 0.07 g cm^{-3} compared to all other mixtures. Although the difference is only 4%, this result is extraordinary, and its explanation is still unknown. The change of ρ by drying and by dehydroxylation is similar to that of the mixtures with FFA and PFA fly-ash. The FBA samples also show a loss of mass by $CaCO_3$ decomposition (see Fig. 5.54). However, the change of ρ measured during cooling (see Fig. 5.55) and after firing at 1100 °C (see Fig. 5.56) are not consistent.

According to Fig. 5.56, the ρ of the FB10 mixture after firing to 1100 °C is higher than the ρ of the reference mixture A0. This is in contrast to what is shown in Fig. 5.55. The explanation for these results is not yet known but is probably again related to the difference in temperature regime (additional 5 min isothermal time at maximum temperature).

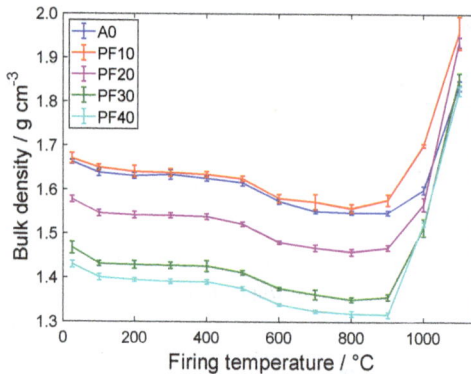

Fig. 5.53 – Bulk density of mixtures with 10 to 40 % PFA and reference mixture A0 after firing at different temperatures.

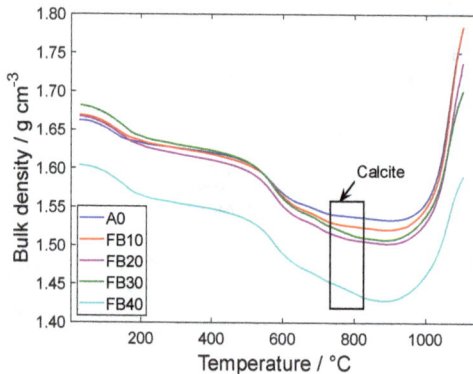

Fig. 5.54 – Bulk density of mixtures with 10 to 40 % FBA and reference mixture A0 during heating.

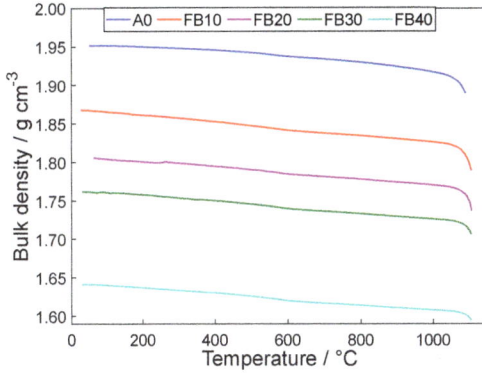

Fig. 5.55 – Bulk density of mixtures with 10 to 40 % FBA and reference mixture A0 during cooling.

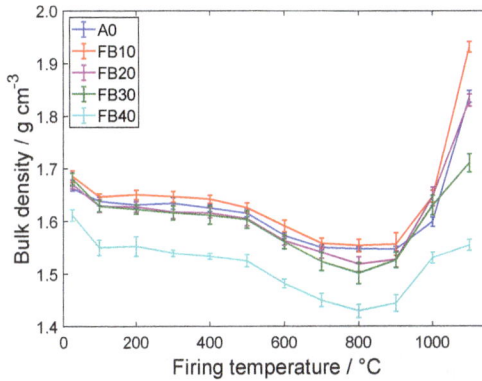

Fig. 5.56 – Bulk density of mixtures with 10 to 40 % FBA and reference mixture A0 after firing at different temperatures.

If a low ρ would be considered a positive property, the most favorable effect is the admixture of FFA fly-ash, even after firing to the highest temperatures. The addition of PFA has a similarly positive effect, but only up to a firing temperature of 900 °C and mixtures containing at least 20 wt.% of PFA. Thus, FBA affects the bulk density only minimally.

5.8   Water absorption and porosity

5.8.1   FFA mixtures

The absorption dependence of the samples from the mixtures of FFA fly-ash and the reference A0 are shown in Fig. 5.57. As can be seen, replacement of a portion of the grog with 10 wt.% FFA does not significantly affect the water absorption. Samples A0 and FF10 have a water absorption of just over 20 % after firing at 800 °C and only about 5 % after firing at 1100 °C. Even the absorption of FF10 is lower than that of the reference mixture A0. With the addition of a further 10 wt.% (FF20) by firing at the highest temperatures, the absorption does not decrease so rapidly and reaches 10 % after firing at 1100 °C. A significant increase in absorption is observed for the FF30 and FF40 mixtures. As can be seen from Fig. 5.58, the water absorption of these mixtures remains practically unchanged with increasing firing temperature, remaining at ~30 % for FF30 and ~35 % for FF40. The decrease in water absorption is closely related to the filling and closing of the pores or the reduction of their diameter. Reducing the diameter of the pores reduces the part of the pores that can be filled with water. This is why the apparent porosity (i.e., the porosity accessible to water) curves in Fig. 5.58 are so similar to the saturation curves. Samples achieve the lowest porosity after firing at 1100 °C from the FF10 mixture (~7%). On the other hand, the FF40 mixture fired at 1100 °C reaches an apparent porosity of up to 45 %. The error bars in the absorption and apparent porosity plots can only be considered as indicative since they were calculated as the standard deviation of the measurements of only three experimental samples.

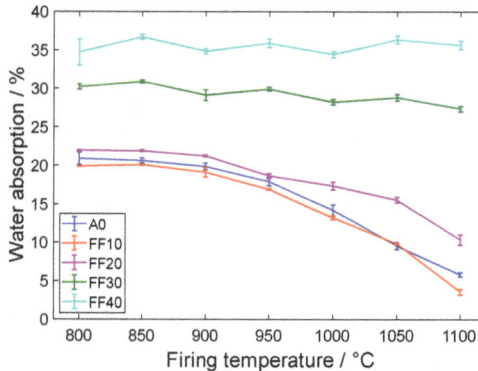

Fig. 5.57 – Water absorption of mixtures with 10 to 40% FFA and reference mixture A0. The error bars represent the standard deviation of the measurement of the three samples.

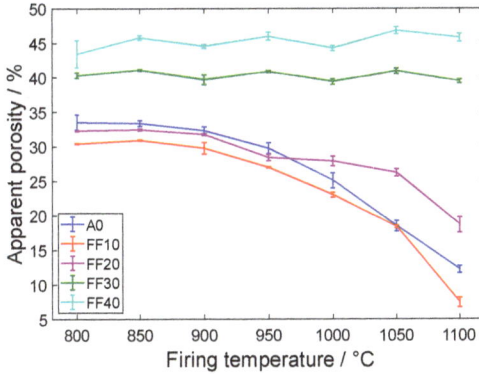

*Fig. 5.58 – Apparent porosity of mixtures with 10 to 40% FFA and reference mixture A0.
The error bars represent the standard deviation of the measurement of the three samples.*

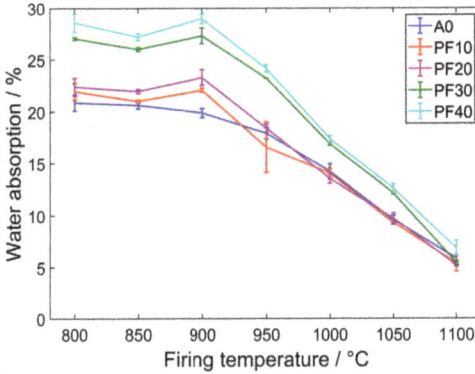

*Fig. 5.59 – Water absorption of mixtures with 10 to 40% PFA and reference mixture A0.
The error bars represent the standard deviation of the measurement of the three samples.*

5.8.2   *PFA mixtures*

The addition of PFA has a more favorable effect on reducing water absorption than FFA.
This can be seen from Fig. 5.59, where the replacement of grog with 10 and 20 wt.% PFA
practically does not change the development of the water absorption of the ceramic bodies
as a function of firing temperature. Even the PF40 mixture, in which PFA fly-ash replaces
all the grog, achieves the same water absorption (~6 %) as the reference A0 mixture after
firing to 1100 °C. The pore volume accessible to water (see Fig. 5.60) again evolves very
similarly to the water absorption (see Fig. 5.59). After firing at 1100 °C, all mixtures
achieve apparent porosities in the range of 10 to 15%.

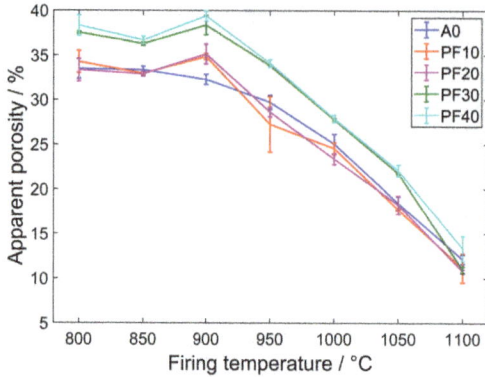

Fig. 5.60 – Apparent porosity of mixtures with 10 to 40% PFA and reference mixture A0. The error bars represent the standard deviation of the measurement of the three samples.

5.8.3 *FBA mixtures*

For mixtures with FBA fly-ash, their dependence of water absorption (see Fig. 5.61) and apparent porosity (see Fig. 5.62) is very similar to the dependence of water absorption for mixtures containing FFA (see Figs. 5.57 and 5.58). Of interest is the mixture of FB10 with 10 wt.% FBA fly-ash. Although FBA is significantly coarser-grained than grog (see results for grain size distribution, Fig. 5.6), the FB10 mixture has both lower water absorption and apparent porosity than the reference sample A0 after firing to all temperatures studied. After firing at 1100 °C, the absorption of the FB10 mixture is only 1.6 %, and the apparent porosity is 3.5 %.

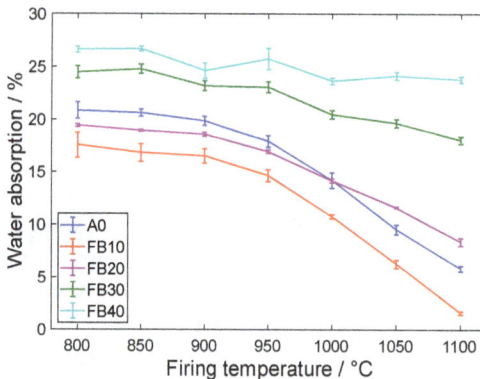

Fig. 5.61 Water absorption of mixtures with 10 to 40% FBA and reference mixture A0. The error bars represent the standard deviation of the measurement of the three samples.

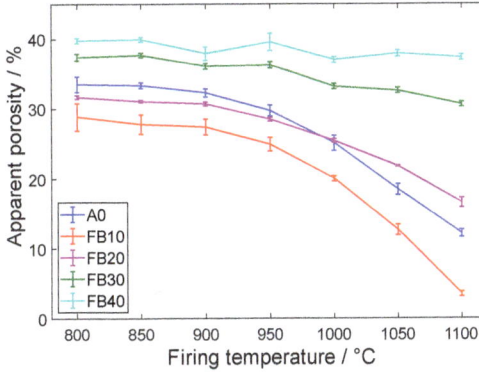

Fig. 5.62 – Apparent porosity of mixtures with 10 to 40% FBA and reference mixture A0. The error bars represent the standard deviation of the measurement of the three samples.

5.8.4   *Pore-size distribution*

Measurements were made solely on A0, FF40, PF40, and FB40 mixtures fired at 900 °C (see Fig. 5.63) and 1100 °C (see Fig. 5.64). After firing at 900 °C, the FB40 mixture has the largest proportion of pores above 1 μm. On the other hand, the A0 and PF40 blends have the most of small pores below 0.1 μm, namely 30%. Firing at 1100 °C generally reduces the volume of the smallest pores. At the same time, the overall porosity is reduced, and thus it can be said that the smallest pores are closed first. This is the most evident in the case of the FB40 blend, where the proportion of pores below 1 μm drops from 58 % to 15 %. An interesting case is the PF40 sample, in which firing at 1100 °C disappeared a significant part of the pores between 0.02 and 1 μm, but the pores below 0.02 μm were preserved. In terms of pore size distribution, FF40 is closest to the reference mixture A0 after firing at 1100 °C. However, the difference is in the total porosity: 5.3 % for A0 compared to 12.1 % for FF40.

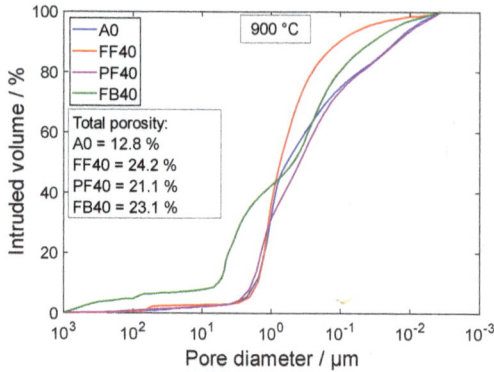

Fig. 5.63 – Cumulative percentage of intruded pore volume of mixtures A0, FF40, PF40, FB40 fired at 900 °C.

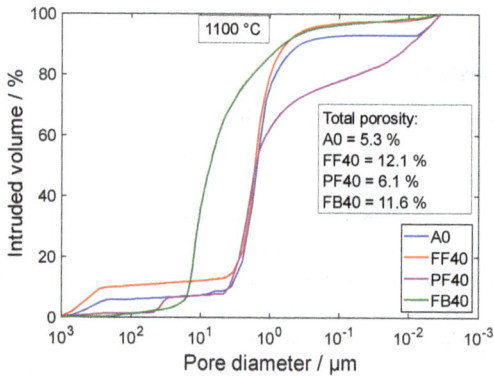

Fig. 5.64 – Cumulative percentage of intruded pore volume of mixtures A0, FF40, PF40, FB40 fired at 1100 °C.

5.9 Thermal diffusivity, thermal conductivity, and specific heat capacity

The results of thermal diffusivity a, thermal conductivity λ, and specific heat capacity at the constant pressure c_p are described together because they were measured by one apparatus. Also, it is functional dependence between them (together with bulk density).

5.9.1 FFA mixtures

The thermal diffusivity a, for the mixtures with FFA as a function of firing temperature is shown in Fig. 5.65. From the qualitative point of view, the effect of FFA on the values of thermal diffusivity is not evident up to the firing temperature of 900 °C. By firing at temperatures up to 300 °C, the thermal diffusivity of all mixtures slightly increases, which is related to the removal of residual moisture. A firing of the mixtures at the temperature of 500 °C and 600 °C, i.e. in the illite dehydroxylation region, causes a decrease in the values of thermal diffusivity of all mixtures by about 15 %. By firing at 1000 °C, thermal diffusivity starts to increase slightly due to the sintering and better heat transfer through the material structure. This behavior is not valid for the FF40 mixture, where the value of thermal diffusivity slightly decreases, which can be explained by the fact that FFA impairs heat transfer. This result is consistent with the increased values of internal friction (see Fig. 5.86), which reflects the extent of energy dissipation of mechanical vibrations in the structure of the material (heat transfer also occurs through vibrations of the atoms of a matter). A significant increase of thermal diffusivity is observed for mixtures of A0, FF10, and FF20 after firing at 1100 °C. The value of thermal diffusivity of the mixture of FF30 increases only slightly, and for the mixture of FF40, it even slightly decreases.

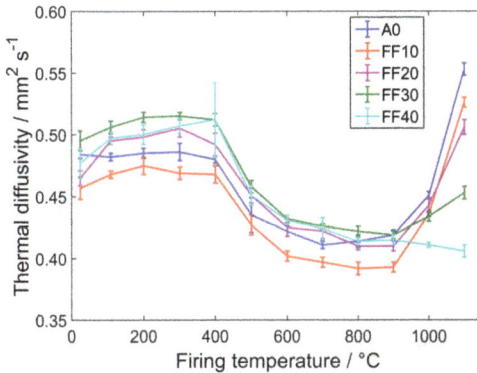

Fig. 5.65 – The thermal diffusivity of mixtures with the
amount of FFA from 0 % to 40 % after firing up to 1100 °C.

The specific heat capacity of the mixtures with FFA is shown in Fig. 5.66. Its values measured for green (dried in the air) samples are inexplicably high, especially for the mixtures of A0, FF10, and FF30. Apparently, there is a connection with the moisture content in the samples, but a clear dependence of c_p on the moisture content is not observed (green samples contained moisture in the order: A0 < FF10 < FF20 < FF30 < FF40). At the moment, these strange results are considered to be due to the unsuitability of the flash comparison method for determining the c_p of wet samples. After drying, i.e., heating to the temperatures of 100 °C and 200 °C, the c_p of the individual mixtures stabilizes at approximately the same values in the range from 1.0 J g^{-1}K^{-1} to 1.5 J g^{-1}K^{-1}. As the firing temperature increases, the differences between the c_p of the individual mixtures are within the measurement uncertainty, and the overall trend can be described as a linear decrease of the c_p with the firing temperature. After firing at 1100 °C, all mixtures have approximately the same mass heat capacity ~0.78 J $g^{-1}K^{-1}$.

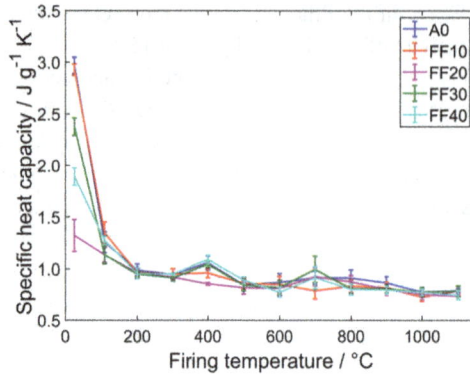

Fig. 5.66 – The specific heat capacity of mixtures with the amount of FFA from 0 % to 40 % after firing up to 1100 °C.

The thermal conductivity (see Fig. 5.67) is calculated from c_p, which is reflected in the values of λ of the green samples. The values of λ decrease up to the firing temperature of 600 °C, at which they stabilize in the range of $0.4 – 0.6$ W $m^{-1}K^{-1}$. It can also be said that the upper limit of this interval is reached for the mixture of A0 and the lower for FF40. For comparison, the values of λ of solid fired bricks are typically in the range of $0.73 – 0.77$ W $m^{-1}K^{-1}$ [104]. These values remain the same for all mixtures up to the firing temperature of 1000 °C. After firing at 1100 °C, the effect of the admixture of FFA ash is manifested, and qualitatively the same conclusions are valid as in the case of thermal diffusivity. The reference mixture A0 reached the value of λ (0.79 ± 0.02) W $m^{-1}K^{-1}$, while the mixture FF40 only (0.40 ± 0.03) W $m^{-1}K^{-1}$. The reduction in thermal conductivity can be considered as a positive effect of replacing traditional grog with FFA ash. However, as mentioned above, the firing of mixtures containing FFA ash at temperatures above 1000 °C causes the release of SO_2, which is unacceptable in the production process.

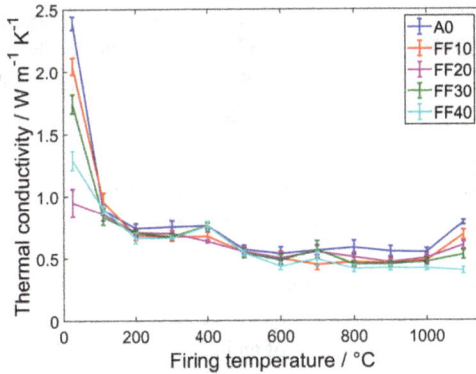

Fig. 5.67 – The thermal conductivity of mixtures with the amount of FFA from 0 % to 40 % after firing up to 1100 °C.

5.9.2 PFA Mixtures

Mixtures containing PFA ash have a lower a as the reference mixture A0 up to the firing temperature of 400 °C and (see Fig. 5.68). As can be seen from the values of a for mixtures containing PFA ash after a firing at the temperatures of 500 °C and 600 °C, they are not significantly affected by the dehydroxylation of illite as it is for the mixture A0. On the other hand, in the area of sintering of a material, i.e. by a firing at the temperatures higher than 900 °C, the increasing of a for the samples containing PFA is practically identical to the sample A0.

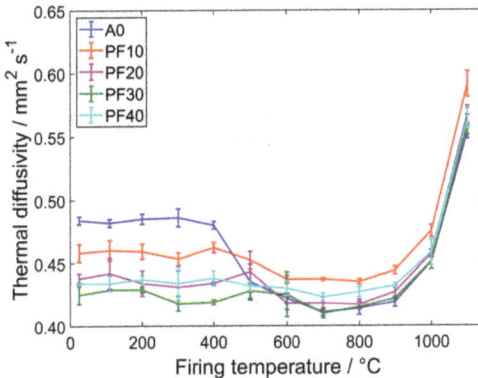

Fig. 5.68 – The thermal diffusivity of mixtures with the amount of PFA from 0 % to 40 % after firing up to 1100 °C.

The mass heat capacity (see Fig. 5.69) of green samples from mixtures containing PFA does not take unreasonably high values as it is for the reference sample A0. After firing at

200 °C, the values of c_p for the mixture A0 are close to the mixtures containing PFA $(0.7 - 1.0$ J g^{-1}K^{-1}). The c_p values are practically not affected by the dehydroxylation process, but they are slightly decreased by firing at the temperature above 800 °C. After firing at 1100 °C, the c_p of all mixtures is practically the same, with the value of (0.75 ± 0.07) J g^{-1}K^{-1}.

The thermal conductivity (see Fig. 5.70) after firing at low temperatures mainly reflects the values of c_p. The thermal conductivity after firing at temperatures above 800 °C is more important for practical applications. As can be seen, the λ values are in the range of $(0.46 - 0.62)$ W m^{-1}K^{-1} after firing at 900 °C and 1000 °C. The smaller values are for the mixtures with a higher PFA content. After firing at the temperature of 1100 °C, the thermal conductivity increases mainly as a result of an increase in thermal diffusivity. The λ values of all mixtures, including the reference A0, are in a narrow range of $0.77 - 0.89$ W m^{-1}K^{-1} after firing at the temperature of 1100 °C, which is a value close to the thermal conductivity for solid fired bricks $0.73 - 0.77$ W m^{-1}K^{-1} [104].

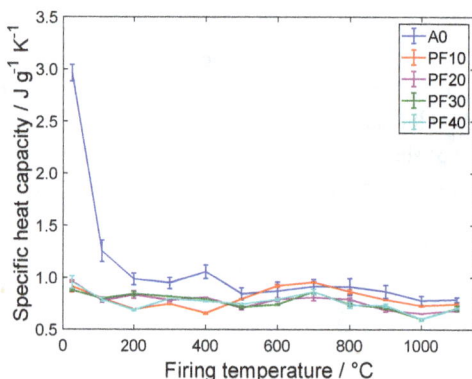

Fig. 5.69 – The specific heat capacity of mixtures with the amount of PFA from 0 % to 40 % after firing up to 1100 °C.

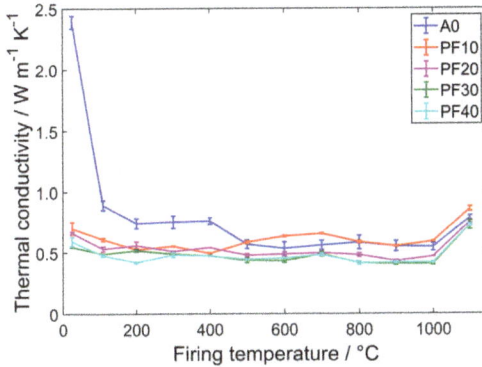

Fig. 5.70 – The thermal conductivity of mixtures with the amount of PFA from 0 % to 40 % after firing up to 1100 °C.

5.9.3   FBA mixtures

The thermal diffusivity of mixtures containing FBA as a function of the firing temperature is shown in Fig. 5.71. Mixtures FF10 and FF30 have a similar trend, and as the reference sample A0. The experimental samples from the mixture FF20 were apparently quite inhomogeneous. Thus the deviation of the measuring values obtained on the individual samples was relatively large, which is reflected in the size of the error bars. The sintering process increases the values of a for all the mixtures. After firing at 1100 °C, the FB10 mixture reaches higher values a than the reference mixture A0. By replacing a grog with FBA ash, it decreases thermal diffusivity, similarly as it was for mixtures with FFA ash (see Fig. 5.65).

The mass heat capacity (see Fig. 5.72) appears to be independent of the amount of FBA. Its values are similar to the mass heat capacity of the FFA mixtures. We do not observe anomalously large c_p values for the green samples from these mixtures, as it was in the case with the mixtures with FFA or with the reference mixture A0. The mass heat capacity of all mixtures after firing at 1100 °C is in the range of $(0.71 - 0.83)$ J $g^{-1}K^{-1}$.

The thermal conductivity is not significantly changed with the firing, and with the increasing of the firing temperature, the differences in thermal conductivity between the mixtures decrease (see Fig. 5.73). After the firing at 1000 °C, the mixtures have values of thermal conductivity in the range of $(0.4 - 0.6)$ W $m^{-1}K^{-1}$. By firing at 1100 °C, the thermal conductivity of all mixtures except FB40 slightly increases. In contrast, the thermal conductivity of the FB40 mixture slightly decreases, similar as it was for mixtures with FFA.

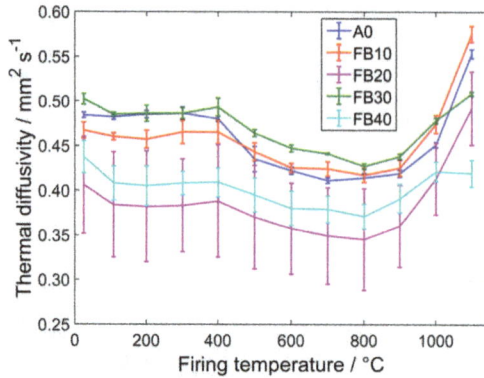

Fig. 5.71 – The thermal diffusivity of mixtures with the amount of FBA from 0 % to 40 % after firing up to 1100 °C.

Low thermal conductivity is a desired property for building materials such as masonry in an exterior part of buildings. The presented results show that thermal conductivity increases mainly by firing at the temperature of 1100 °C. Therefore, from the point of view of maintaining a low thermal conductivity, it would be advantageous to apply firing temperatures lower than 1100 °C. However, the aspect of sufficient mechanical strength must also be considered. Mechanical strength increases by firing at the highest temperatures (see chapter 5.12). Therefore, a compromise between mechanical strength and thermal conductivity with respect to the specific application of the material must be found.

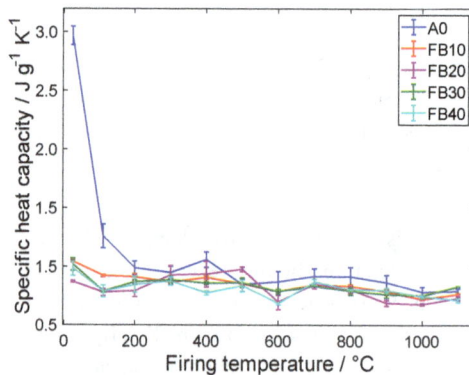

Fig. 5.72 – The specific heat capacity of mixtures with the amount of FBA from 0 % to 40 % after firing up to 1100 °C.

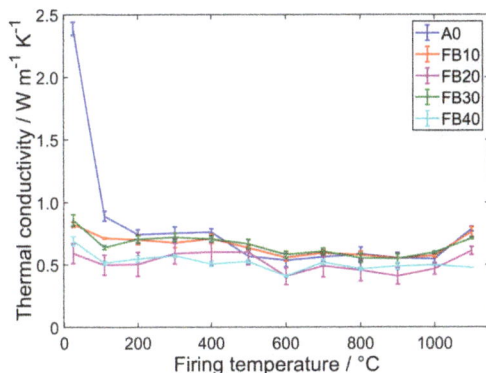

Fig. 5.73 – The thermal conductivity of mixtures with the amount of FBA from 0 % to 40 % after firing up to 1100 °C.

5.9.4   Thermal diffusivity during a heating

Measurement during heating was performed only on reference sample A0 and samples FF40, PF40, FB40, where all classical grog was replaced by ash. The results are in Figs. 5.74 – 5.77. In the figures gray dots represent the measured values at a given temperature, and the black dots are their arithmetic mean. Thermal diffusivity decreases up to 550 °C of all mixtures, as it is typical for the most of materials. At a temperature around 600 °C thermal diffusivity of all mixtures increased. According to the temperature, this increase can be attributed to the dehydroxylation of illite. Further development is difficult to interpret. In the temperature range of 800 – 900 °C, a slight decrease is observed, which could be attributed to the complete disintegration of the illite structure. Here, the decomposition of $CaCO_3$ in FFA and FBA ashes could also have a certain effect. Next, as the temperature increases thermal diffusivity also increases as a result of sintering. In the case of replacement of grog with FFA ash, this increase is significantly reduced. Similar results were also found during measurements of thermal diffusivity after a firing.

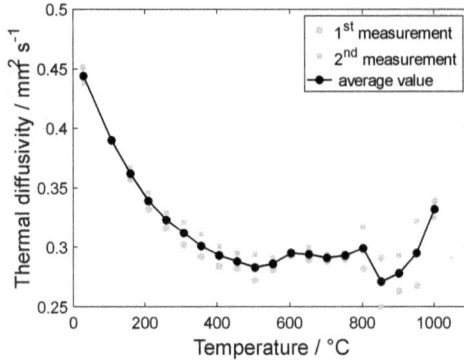

Fig. 5.74 – Thermal diffusivity of mixture A0 during quasi-linear heating with a rate of 3.3 °C min⁻¹. Dynamic atmosphere N_2 100 ml min⁻¹. Gray dots represent the measured values at a given temperature, and the black dots are their arithmetic mean.

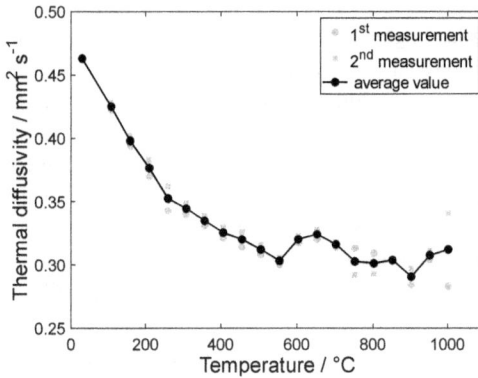

Fig. 5.75 – Thermal diffusivity of mixture FF40 during quasi linear heating with a rate of 3.3 °C min⁻¹. Dynamic atmosphere N_2 100 ml min⁻¹. Gray dots represent the measured values at a given temperature, and the black dots are their arithmetic mean.

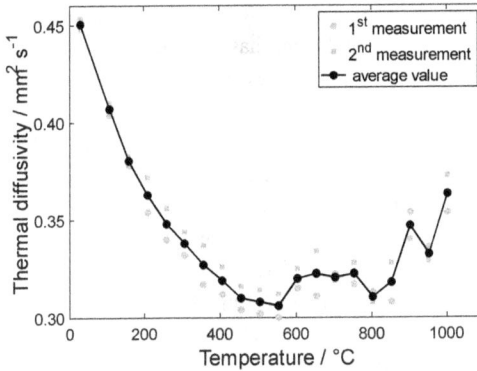

Fig. 5.76 – Thermal diffusivity of mixture PF40 during quasi linear heating with a rate of 3.3 °C min⁻¹. Dynamic atmosphere N_2 100 ml min⁻¹. Gray dots represent the measured values at a given temperature, and the black dots are their arithmetic mean.

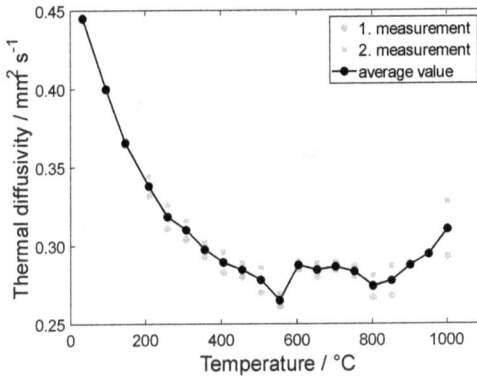

Fig. 5.77 – Thermal diffusivity of mixture FB40 during quasi linear heating with a rate of 3.3 °C min⁻¹. Dynamic atmosphere N_2 100 ml min⁻¹. Gray dots represent the measured values at a given temperature, and the black dots are their arithmetic mean.

5.10 Young's modulus

5.10.1 FFA mixtures

Young's modulus of the mixtures containing FFA ash and reference mixture A0 during the heating and cooling is shown in Fig. 5.78. The released of physically bound water in the A0 mixture caused a significant increase of Young's modulus from an initial value of 3.67 GPa to 7.39 GPa at 279 °C (this increase is about 100%). The addition of FFA instead of grog reduces this increase, even though mixtures with FFA contain more physically bound water. Moreover, it can be seen that the increase of the values of Young's modulus during released of physically bound water has two rates, first up to a temperature of ~120 °C and second up to ~240 °C. The values of Young's modulus increase in spite of a decrease of bulk density (direct proportionality between them is valid) and a slight increase of dimensions (indirect proportionality between them is valid). Such behavior is typical for materials containing kaolinite and/or illite if the samples are not completely dried [105, 106]. The increase of Young's modulus is due to the removal of physically bound water from the pores between the material particles. Then, these particles can make closer contact with each other due to van der Waals forces. The temperature where Young's modulus reached its maximum after drying is interesting. This temperature does not correlate with the release of physically bound water, which ends at a temperature of ~180 °C (see Fig. 5.22). This phenomenon is not explained yet. After removal of physically bound water (after drying), Young's modulus starts to decrease with the temperature, as it expected for standard materials. This decrease is slightly accelerated by the dehydroxylation of illite.

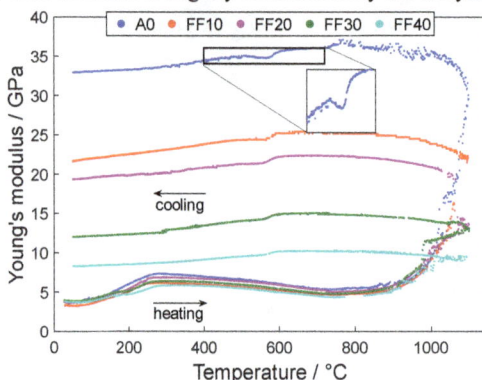

Fig. 5.78 – Young's modulus of the mixtures with FFA addition from 10 wt.% to 40 wt.% and the reference mixture A0 during a firing.

From ~750 °C, Young's modulus begins to increase slowly, which can be attributed to the start of solid-phase sintering. For the mixture of A0 and mixtures with a lower FFA content (FF10 and FF20) is valid that Young's modulus rises with increasing temperature as a result of melting and viscous flow sintering. The presence of a liquid phase (or low viscosity glassy phase) also causes problems in measuring the resonant frequency. The sample obtains pyro plastic properties, and the mechanical vibration is damped very quickly (this is also reflected in the measurement of the logarithmic decrement Fig. 5.89). This is the reason for the scattering of Young's modulus at high temperatures. Viscous flow sintering in the case of mixtures with a higher field content (FF30 and FF40) is suppressed, and only a markedly increase of Young's modulus in the temperature range between 900 °C and 1000 °C is observed. This can be attributed to the formation of the anorthite, which was visible on the DTA curves (see Fig. 5.21).

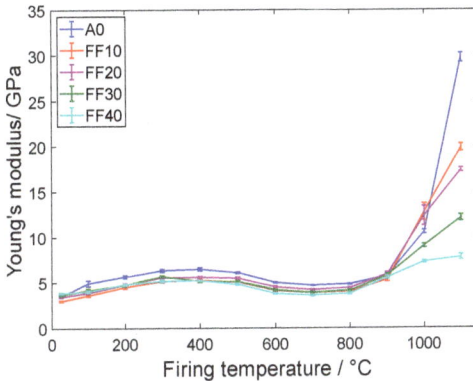

Fig. 5.79 – Young's modulus of the mixtures with FFA addition from 10 wt.% to 40 wt.% and the reference mixture A0 after a firing at different temperatures.

The addition of FFA causes a relatively significant reduction of Young's modulus, as can be seen from its values during cooling. Nevertheless, it is also visible from measurements after firing (see Fig. 5.79). The increasing of Young's modulus continues even after the start of the cooling stage of a firing as a result of the continued sintering and the increase in the viscosity of the glass phase. This trend continues until the temperature of a glass transition (~800 °C) is reached. Below this temperature, the glass phase can no longer absorb stresses that originate in the different thermal expansion of the glass phase and the quartz (and probably other solid phases), and microcracks start to form. During cooling, a modification β-α transformation occurs in the quartz crystals, in which the quartz reduces its volume by 0.68% [107]. This causes the circumferential cracks formation reflected in Young's modulus as a small drop (minimum) at 573 °C [39]. This effect is less visible in the case of mixtures FF30 and FF40 with a high content of FFA ash as a consequence of the lower amount of glassy phase. When the cracks around the quartz grains are partially closed (due to the more significant thermal expansion of the glass matrix), the decrease of Young's modulus continues as a result of the formation of further cracks due to different thermal expansion of the present phases. Crack formation at temperatures below 573 °C in illite-based ceramic bodies was confirmed by acoustic emission measurements[39]. Measurements of Young's modulus during firing and after firing are in good agreement, although at 1100 °C, the measurements after firing show about 10% lower values than the measurements during firing. The reason is unknown, but more worthy are the measurements after firing (see Fig. 5.79).

5.10.2 PFA mixtures

The results of Young's modulus of samples with PFA ash during a firing are in Fig. 5.80 (Important firing areas, which are reflected in Young's modulus are marked). Qualitatively, these results are similar to those for mixtures with FFA ash. It is interesting that the increase of Young's modulus after drying is approximately the same as for the mixtures with FFA, although the mixtures with PFA had lower initial moisture (PF40 had initial moisture ~1.3% and FF40 ~3.4%). The main differences are for Young's modulus of the mixtures after firing. While Young's modulus was significantly reduced when grog was replaced with FFA ash, PFA ash had only little effect on the values of Young's modulus. Even mixtures containing PFA after firing at 1000 °C and 1100 °C achieve higher values of Young's modulus than the reference sample A0 Fig. 5.81). Interestingly, during cooling below the temperature of quartz transition, Young's modulus of the mixtures with PFA decreases with a temperature steeper than Young's modulus of the reference mixture A0 (see Fig. 5.80), which indicates an increased formation of microcracks. The measurements of Young's modulus indicate that PFA can be used as a suitable replacement of conventional grog.

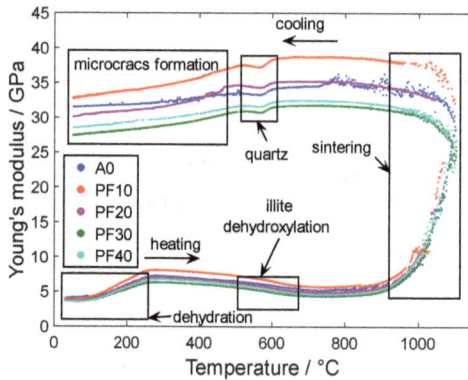

Fig. 5.80 – Young's modulus of the mixtures with PFA addition from 10 wt.% to 40 wt.% and the reference mixture A0 during a firing.

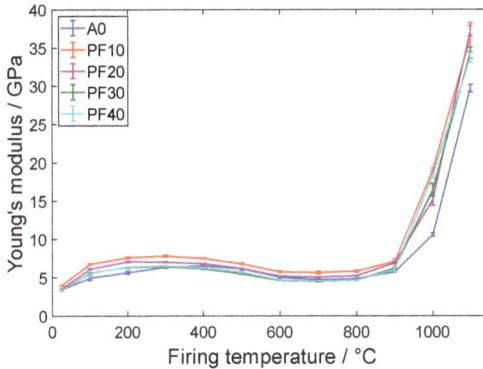

Fig. 5.81 – Young's modulus of the mixtures with PFA addition from 10 wt.% to 40 wt.% and the reference mixture A0 after a firing at different temperatures.

5.10.3 FBA mixtures

The temperature dependence of Young's modulus of the mixtures with FBA during heating (see Fig. 5.82) and after a firing (see Fig. 5.83) is quantitatively similar to the previous mixtures (the mixtures with FFA or PFA). Nevertheless, it is visible that the addition of FBA dramatically reduces Young's modulus. However, in the case of complete replacement of grog with FBA ash (the mixture FB40), Young's modulus after a firing at 1100 °C is higher than in the case of complete replacement of grog with FFA ash (the mixture FF40 in Fig. 5.78).

The decrease of Young's modulus is very significant during the quartz transition in the cooling stage of a firing. FBA contains up to 44 wt.% of quartz (see Table 5.4).

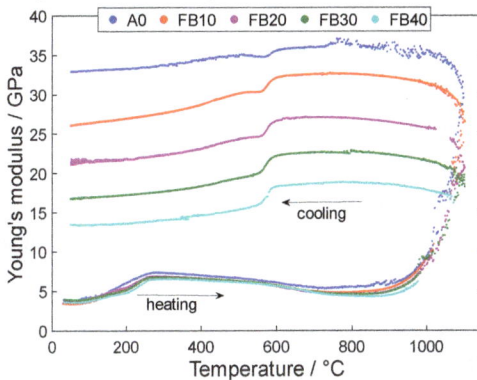

Fig. 5.82 – Young's modulus of the mixtures with FBA addition from 10 wt.% to 40 wt.% and the reference mixture A0 during a firing.

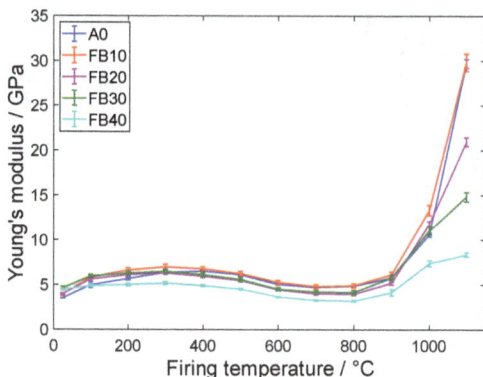

*Fig. 5.83 – Young's modulus of the mixtures with FBA addition from 10 wt.% to 40 wt.%
and the reference mixture A0 after a firing at different temperatures*

5.11 Internal friction

The internal friction, which is represented in this work by the logarithmic decrement (δ), can be considered as a measure of the dissipation of energy of mechanical vibrations on structural defects in the material [108]. The measurement of the logarithmic decrement is associated with a relatively large scatter. The presented curves of logarithmic decrement during a firing were smoothed by the local fitting of 20 points with a quadratic function.

5.11.1 FFA mixtures

The evolution of the logarithmic decrement for mixtures containing FFA during heating is shown in Fig. 5.84. No significant differences are observed between the individual mixtures. At the beginning of the heating, the logarithmic decrement of all mixtures decreases as a result of the drying process. Thus, it can be concluded that the water adsorbed on the surface of the particles or water in pores dissipates the energy of mechanical vibrations. The minimum value of the logarithmic decrement is at a temperature of ~400 °C. Subsequently, the logarithmic decrement increases due to the dehydroxylation of illite. Then, a slight decrease of the logarithmic decrement is visible from the temperature of ~700 °C, which can be attributed to solid-state sintering. The formation of the glassy phase causes an increase of the logarithmic decrement (from ~890 °C). When the maximum temperature of 1100 °C is reached, the cooling stage starts (see Fig. 5.85). The logarithmic decrement of the FFA mixtures decreases with temperature exponentially as a result of the decrease of the viscosity of the glassy phase. A different course is observed in the case of reference mixture A0, where a sharp decrease of the logarithmic decrement occurs in the temperature interval from 800 °C to 700 °C. This could be explained by a glass transition, but it could be only a measurement error. The quartz transition at ~573 °C creates cracks i.e., defects, which are also reflected in the values of logarithmic decrement. During further cooling below 573 °C, logarithmic decrement increases slightly with a maximum at a temperature of ~300 °C and then decreases again. Such a development is in accordance with the hypothesis of crack formation.

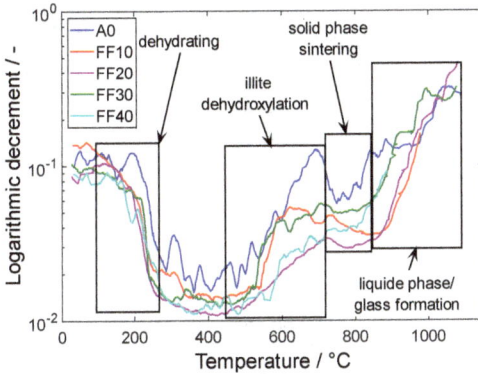

Fig. 5.84 – The logarithmic decrement of the mixtures with FFA addition from 10 wt.% to 40 wt.% and the reference mixture A0 during a heating.

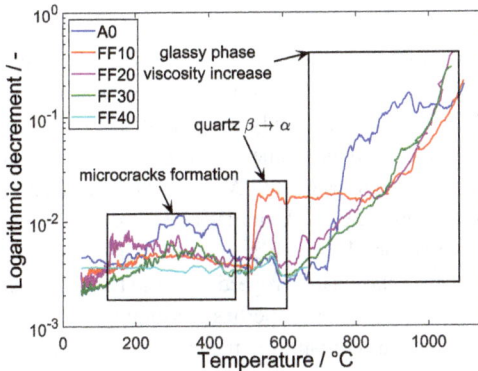

Fig. 5.85 – The logarithmic decrement of the mixtures with FFA addition from 10 wt.% to 40 wt.% and the reference mixture A0 during a cooling.

Measurements of the logarithmic decrement after firing mixtures containing FFA at different temperatures (see Fig. 5.86) show its exponential decrease with a firing temperature. The addition of 10 wt.% and 20 wt.% of FFA results in slightly lower values of the logarithmic decrement compared to the reference mixture of A0 after firing at 1100 °C, while the addition of 30 wt.% and 40wt.% of FFA has the opposite effect. It can be said that the structure of mixtures containing 30 wt.% and 40wt.% of FFA is more failure compared to the reference mixture A0. The significant quantitative differences between the logarithmic decrement measured during firing and after firing are not yet explained.

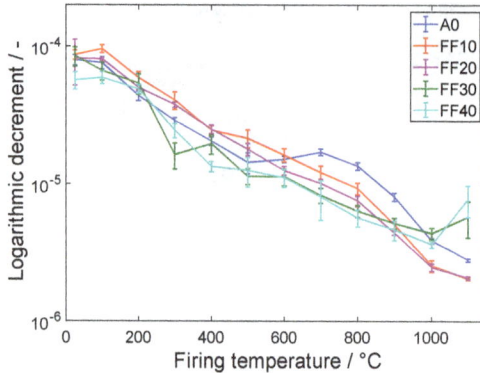

Fig. 5.86 – The logarithmic decrement of the mixtures with FFA addition from 10 wt.% to 40 wt.% and the reference mixture A0 after a firing at different temperatures.

5.11.2 PFA mixtures

The results of the logarithmic decrement during a firing of mixtures with PFA (see Fig. 5.87) are essentially identical to the results for mixtures with FFA. The difference is during cooling (see Fig. 5.88), where the peak at 573 °C, which is associated with quartz transition, is not observed because PFA contains only 3 wt.% of quartz. The observation that the decrease of the logarithmic decrement due to the drying process (see Fig. 5.87) is as significant as in the case for mixtures with FFA, in spite of the fact that mixtures with PFA contained less moisture, is interesting. A similar phenomenon was observed for Young's modulus. This can be explained by the fact that only water molecules adsorbed on the surface of the particles affect the elastic properties, while the water molecules in the pores do not affect the elastic and inelastic properties. Assuming that the water molecules are released from the open pores faster than from the interparticle space, this may also explain the shift of the increasing of Young's modulus and the decreasing of the logarithmic decrement to the higher temperatures.

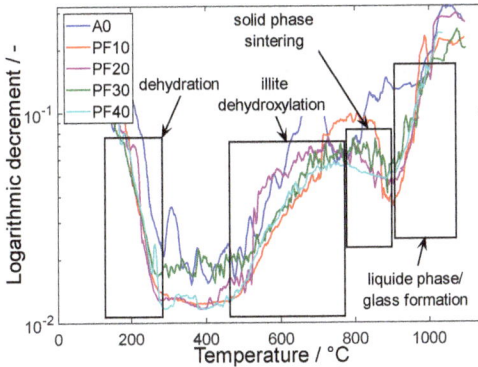

Fig. 5.87 – The logarithmic decrement of the mixtures with PFA addition from 10 wt.% to 40 wt.% and the reference mixture A0 during a heating.

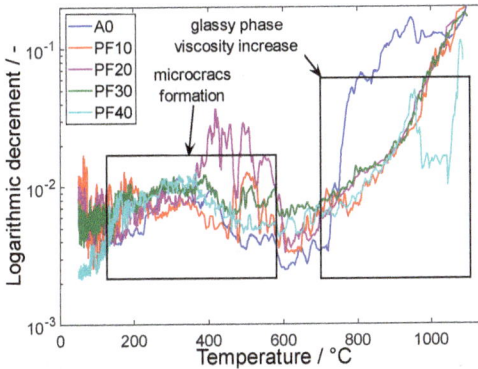

Fig. 5.88 – The logarithmic decrement of the mixtures with PFA addition from 10 wt.% to 40 wt.% and the reference mixture A0 during a cooling.

The measurements of the logarithmic decrement after a firing (see Fig. 5.89) show that internal damping of mixtures with PFA is smaller, which leads to fewer failures. No specific trend of the logarithmic decrement as a function of PFA content is observed. It decreases exponentially up to a firing temperature of 300 °C. After firing at (400 – 700) °C, the logarithmic decrement is almost constant. At temperatures above 700 °C, the logarithmic decrement starts to decrease again due to the arrangement of the structure due to sintering.

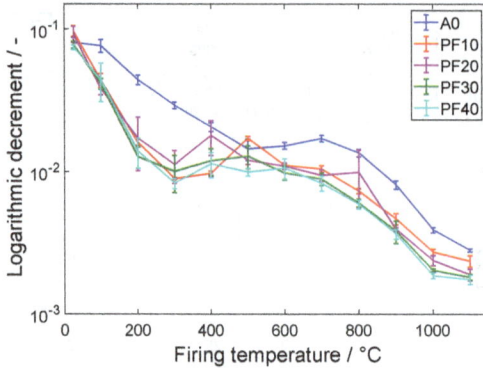

Fig. 5.89 – The logarithmic decrement of the mixtures with PFA addition from 10 wt.% to 40 wt.% and the reference mixture A0 after a firing at different temperatures.

5.11.3 FBA mixtures

The same conclusions about the logarithmic decrement development are also valid for the mixtures with FBA during heating (see Fig. 5.90) and cooling (see Fig. 5.91) as they were for mixtures with FFA and PFA. Due to a large amount of quartz in the FBA (44 wt.%), it would be logical to expect its effect on the development of the logarithmic decrement during cooling. However, as shown from Fig. 5.91, no peak is observed around 573 °C, as it was in the case of the mixtures with FFA (see Fig. 5.85). The large scatter of experimental points of the logarithmic decrement during the cooling stage is probably the reason.

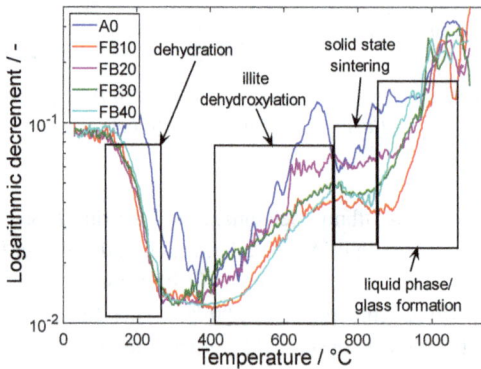

Fig. 5.90 – The logarithmic decrement of the mixtures with FBA addition from 10 wt.% to 40 wt.% and the reference mixture A0 during a heating.

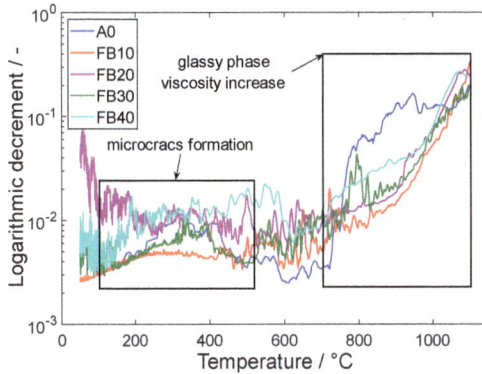

Fig. 5.91 – The logarithmic decrement of the mixtures with FBA addition from 10 wt.% to 40 wt.% and the reference mixture A0 during a cooling.

The results after firing at temperatures up to 1000 °C show that the mixtures with FBA reach the lowest values of the logarithmic decrement of all investigated mixtures (see Fig. 5.92). The quartz, which is part of FBA (about 44 wt.%), is in the mixture in the form of relatively large grains (see results of grain distribution in Fig. 5.89 or micrograph in Fig. 5.86), and therefore there is a relatively small number of defects in mixture volume. This is reflected in low values of the logarithmic decrement. Next, during increasing the firing temperature, a melt is formed, and sintering with a viscous flow occurs, which after firing reduces the logarithmic decrement. However, at the same time, more cracks are formed during cooling due to the quartz transition in the boundary of quartz and melt, which on the other hand, contributes to the increase of the logarithmic decrement. As the glass-quartz interface increases with a firing at higher temperatures, the contribution of cracks around the quartz grains to the logarithmic decrement is also greater after a firing at higher temperatures.

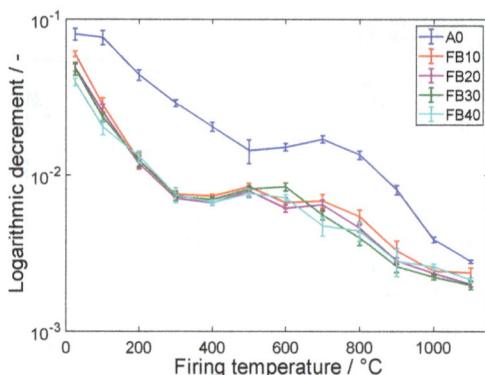

Fig. 5.92 – The logarithmic decrement of the mixtures with FBA addition from 10 wt.% to 40 wt.% and the reference mixture A0 after a firing at different temperatures.

5.12 Mechanical strength

The mechanical strength was measured on dried samples at 110 °C, and after a firing at 800 °C, 1000 °C, and 1100 °C using three-point bending.

5.12.1 FFA mixtures

The results of the mechanical strength of mixtures containing FFA are shown in Fig. 5.93. Immediately after the drying process, the mixture A0 reaches values of a mechanical strength (~8 MPa), which is almost twice compared to mixtures containing FFA (~4 MPa). The mechanical strength of ceramic bodies before a firing is important in the production process due to the safety manipulation with products. In this respect, the addition of FFA has a negative effect. Nevertheless, the condition of minimum bending strength for raw ceramic tiles of 2 MPa [12] is fulfilled. After a firing at 800 °C, the mechanical strength of the mixture A0 dropped down to (6.4±0.7) MPa, while the mechanical strength of the mixture with FFA increases to 55 MPa. The decrease of mechanical strength of the A0 mixture could be explained by the dehydroxylation of illite, but the reason for the increase in the case of mixtures with FFA is still unclear. After firing at 1000 °C and 1100 °C, the mechanical strength increases, but this increase decreases with the addition of FFA. This development is similar to that of the Young's modulus (see Fig. 5.79), which confirms the rule of direct proportionality between the Young's modulus and mechanical strength (see chapter 4.13). The values of the mechanical strength of all studied mixtures after a firing at 1000 °C and 1100 °C meet the requirements of class AIII according to norm BS EN 14411 [109] (see Table 4.5).

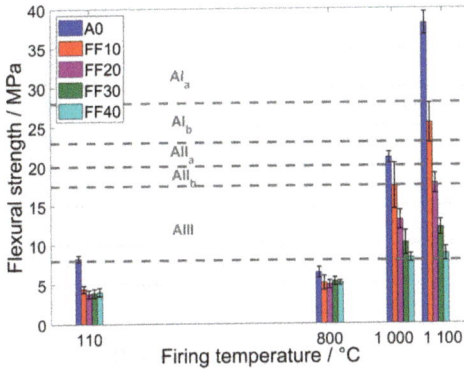

Fig. 5.93 – The mechanical strength of the mixtures with FFA addition from 10 wt.% to 40 wt.% and the reference mixture A0 after a firing at different temperatures.

5.12.2 PFA mixtures

As it was for FFA, also PFA reduces the mechanical strength of green bodies (temperature 110 °C in Fig. 5.94). The mixture PF40 has the mechanical strength (2.82 ± 0.20) MPa after drying, therefore it still meets the minimum requirement of strength for safe manipulation, which is 2 MPa. Again, an interesting phenomenon with values of the mechanical strength can be seen. Here, as it was for mixtures with different fly ash, the mechanical strength of mixtures containing PFA increases after a firing at a temperature of 800 °C, while the mechanical strength of reference mixtures A0 without ash decreases. By firing at 800 °C, no mixture meets the requirements of class AIII for the production of extruded ceramic tiles. A significant increase of the mechanical strength occurs by a firing at a temperature of 1000 °C. In this case, mixtures with PFA ash have better mechanical strength than mixtures with FFA. As it can be seen, the mechanical strength decreases with an increasing amount of PFA. However, this decrease is not so dramatic as it was for mixtures with FFA. Mixtures PF10 and PF20 achieve the mechanical strength required for class AIIb, which includes extruded ceramic tiles. Mixtures PF30 and PF40 fall into the class AIII without any problems. During the next increase of the firing temperature up to 1100 °C, the mechanical strength of all mixtures increases. The slightest mechanical strength (22.82 ± 0.93) MPa is for PF40, but it can still be classified as class AI_b. There is no clear correlation between the mechanical strength, Young's modulus, and fly ash content, as it was in the case of mixtures with FFA. Although Young's modulus of mixtures containing PFA reaches higher values than reference mixture A0 after a firing at temperatures of 1000 °C and 1100 °C, in the case of the mechanical strength, the situation is the opposite, and with an increasing proportion of PFA, the mechanical strength decreases.

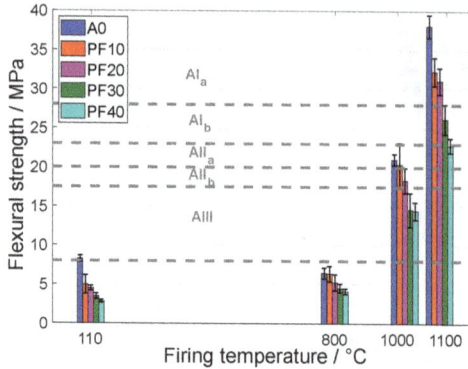

Fig. 5.94 – The mechanical strength of the mixtures with PFA addition from 10 wt.% to 40 wt.% and the reference mixture A0 after a firing at different temperatures

5.12.2 FBA mixtures

The mechanical strength of mixtures containing FBA after a firing at different temperatures is shown in Fig. 5.95. The dried samples (110 °C) reach the same mechanical strength as for the mixtures with FFA. By firing at 800 °C, the mechanical strength of the mixtures with FBA does not increase as it was in the previous two used ashes. After a firing at 1000 °C, all mixtures can be classified in the class AIII, although it can be seen that the mechanical strength is less than for mixtures with FFA and PFA. A similar statement is valid even after a firing at a temperature of 1100 °C. In this case, only mixture FB10 falls into the class AIb, and other mixtures can be classified in the class AIII.

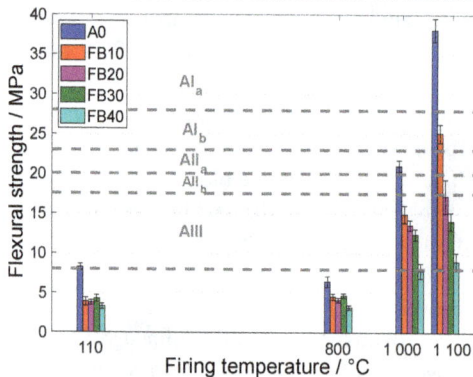

Fig. 5.95 – The mechanical strength of the mixtures with FBA addition from 10 wt.% to 40 wt.% and the reference mixture A0 after a firing at different temperatures.

5.13 The study of microstructure

The microstructure is studied on the basis of micrographs from scanning electron microscopy (SEM). The images were taken on the fracture surfaces of the samples fired at a temperature of 1100 °C.

The micrographs of the fracture surfaces of the sample from the mixture A0 fired at 1100 °C are shown in Fig. 5.96. The original illite crystals were melted into the form of a glass. This can be seen from the rounded shapes and relatively smooth edges. The structure is relatively compact with low porosity. All parts of the structure are interconnected by the glass and invisibly isolated, unreinforced grains. This is consistent with the high mechanical strength of the A0 mixture after firing at 1100 °C.

Fig. 5.96 – SEM micrographs of the fracture surfaces of the sample from the mixture A0 fired at 1100 °C with a soaking time of 5 min.

In contrast, to reference mixture A0, only a small amount of glass was formed in the mixture FF40 (see Fig. 5.97) by firing at 1100 °C. The structure is much more porous. The original illite crystals only partially bonded to each other.

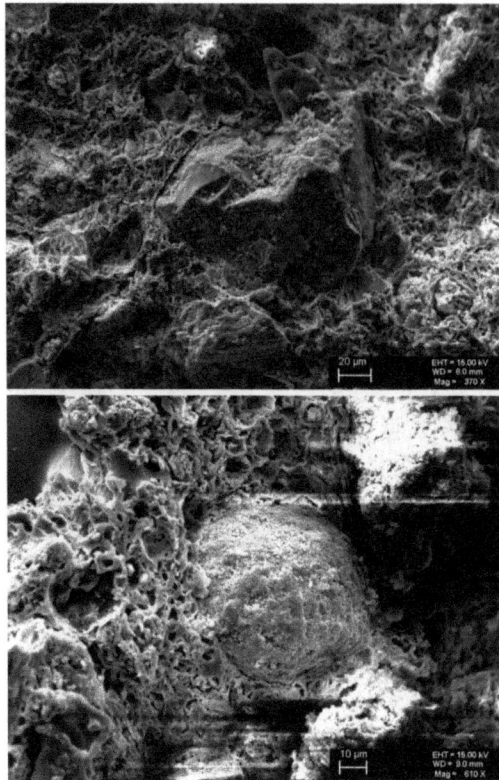

Fig. 5.97 – SEM micrographs of the fracture surfaces of the sample from the mixture FF40 fired at 1100 °C with a soaking time of 5 min.

A much better degree of sintering can be deduced from the micrographs of the surface fracture of mixture PF40 fired at 1100 °C (see Fig. 5.98). Many predominantly circular pores can be seen in the glass matrix. An increased proportion of closed porosity can be expected. Illite and PFA have the ability to work together to create a compact, uniform structure. The mechanical properties can be impaired by a number of pores, which, on the other hand, slightly reduce the thermal conductivity, as shown in Fig. 5.70.

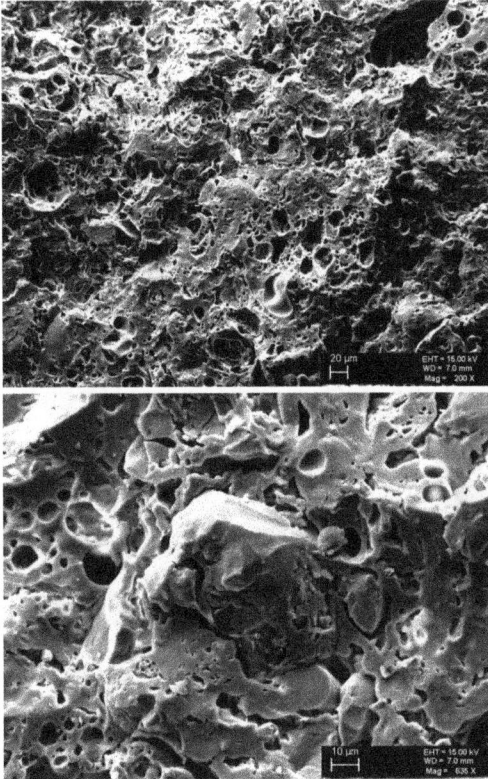

Fig. 5.98 – SEM micrographs of the fracture surfaces of the sample from the mixture PF40 fired at 1100 °C with a soaking time of 5 min.

Looking at the surface fracture of the mixture FB40 fired at 1100 °C (see Fig. 5.99), the high quartz content is confirmed. The quartz grains are embedded in a matrix that appears to be largely made of glass. The grain boundaries look to be relatively weak, resulting in low measured values of mechanical strength.

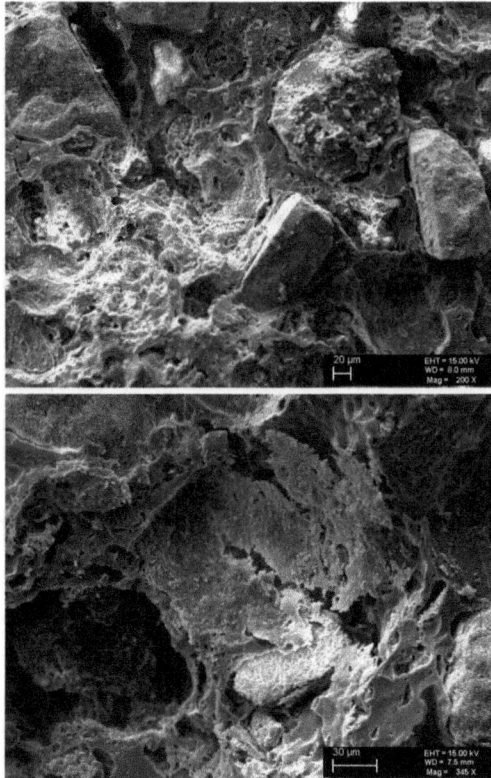

Fig. 5.99 – SEM micrographs of the fracture surfaces of the sample from the mixture FB40 fired at 1100 °C with a soaking time of 5 min.

Conclusions

In this monograph, a comprehensive study of the effect of admixture of power plant fly ash from lignite combustion on the properties of ceramic bodies was presented. Attention was paid in particular to the properties related to the products of building ceramics. A systematic procedure for replacing the classical grog used in the production of traditional ceramics with fly ash was chosen. Subsequently, the effect of a grog replacement on the drying process, weight loss, and volume changes by firing on the bulk density, absorption, porosity, pore distribution, thermal properties, Young's modulus, internal damping, flexural strength, microstructure, and phase composition was investigated. Three types of fly ash from the Nováky thermal power plant were investigated: FFA - fly ash from the fluidized bed boiler; PFA - fly ash of fine fraction from pulverized coal combustion boilers; and FBA - fluidized bed combustion bottom ash. The mixtures were produced by wet mixing; i.e., mixing the raw materials in the required ratio with water and subsequent shaping of the samples. Most of the samples were produced by plastic extrusion. All the experimental mixtures contained 60 wt.% of clay soil with a high content of the mineral called illite which represented the plastic component of the ceramic mixture. The remainder was composed of grog which was replaced by fly ash in varying amounts. Many material parameters were measured after firing to different temperatures and in-situ during the firing process.

The following findings were obtained for the substitution of FFA grog by fly ash:

- It reduces overall shrinkage by firing. When grog was completely replaced with FFA fly ash, the reduction in burn shrinkage was from 4.49 % to 0.69 %. In addition, sintering shrinkage takes place in one extra step, already at 889 °C as a result of anorthite formation;
- It reduces the bulk density. The lower bulk density is maintained throughout the firing process. After firing at 1100 °C, the mixture with 40 wt.% of FFA has 28 % lower bulk density than the mixture without fly ash;
- It reduces the thermal conductivity. This property is not essential for tiles but is beneficial when mixtures with FFA are used for the production of building envelope masonry elements. The thermal conductivity of the mixture with 40 wt.% of FFA after firing to 1100 °C is 0.40 W $m^{-1}K^{-1}$, which is 49 % less than the thermal conductivity of the mixture without fly ash;
- It promotes the formation of anorthite. Anorthite is considered to be a mineral that increases the strength of ceramic materials and reduces shrinkage by firing;
- It increases water absorption. Increased water absorption usually reduces frost resistance and generally accelerates the degradation of the properties of the ceramic body;

- It decreases Young's modulus of elasticity. Its measurements showed that FFA suppresses sintering by viscous flow and promotes improved mechanical properties due to increased anorthite formation. Overall, however, by firing to 1100 °C, Young's modulus decreases steeply with the addition of FFA. A linear relationship between the decrease in Young's modulus and flexural strength was observed;

- It decreases the flexural strength. If all the grog is replaced by FFA fly ash, the flexural strength does not increase further after firing to temperatures above 1000 °C. After firing to 1000 °C, all mixtures meet the minimum requirements of Class AIII for the production of ceramic tiles by extrusion. By replacing all the FFA grog with fly ash, the flexural strength after firing at 1000 °C decreased from (21.0 ± 0.7) MPa to (8.3 ± 0.5) MPa, a decrease of 60 %;

- Sulfur oxides are released from FFA at temperatures approaching 1100 °C. This behavior is unacceptable for modern manufacturing processes and thus limits the maximum firing temperature;

- It increases internal damping. Greater internal damping indicates a more defective structure. However, such material may exhibit better sound insulation properties;

- It increases the porosity. Increased porosity, on the one hand, improves thermal insulation properties and lightens the overall material, but on the other hand, increases water absorption with reduced mechanical strength;

- It does not affect the pore size distribution after firing at 1100 °C.

- To improve the mechanical properties of FFA-containing blends, research attention should focus on the anorthite formation process.

Addition of PFA fly ash:

- PFA has a chemical composition very close to the primary plastic component.

- It reduces the bulk density up to a firing temperature of 1000 °C. After firing above 1000 °C, however, it becomes equal in bulk density to the reference mixture without fly ash.

- Compared to other fly ashes, it has the best sinterability in combination with illite.

- It reduces water absorption. After firing at 1100 °C, the water absorption of the PFA-containing blends was close to that of the reference blend without fly ash (5 – 7 %).

- It increases shrinkage by firing.

- It increases microcracking on cooling below 570 °C.

- It decreases flexural strength. However, it achieves the best flexural strength results compared to other fly ashes. After firing at 1000 °C, the flexural strength of the grog-free blend, i.e., PF40, is (14.4 ± 1.0) MPa, which is 31 % lower than the flexural strength of the reference A0 blend without fly ash.

- It increases Young's modulus of elasticity. On the one hand, a higher Young's modulus indicates better mechanical properties; on the other hand, for the same mechanical strength, a lower Young's modulus is preferable because the material can withstand a higher deformation.
- It has little effect on water absorption and apparent porosity.
- Firing at 1100 °C preserves pores below 0.02 μm in PFA-containing bodies, which account for up to ~20% of the total porosity.
- Little effect on thermal conductivity.
- Overall, PFA has the least impact on the material's properties in which it replaced grog.

The admixture of FBA has the following impacts:

- A reduction of water absorption with a maximum addition of 10 wt.% FBA.
- More large pores after firing to 1100 °C. Up to 85 % of the pores are above 1 μm in size. Larger pores often have a beneficial effect on frost resistance.
- A reduction in the flexural strength. After firing at 1000 °C, the flexural strength of the mixture where all the grog has been replaced by FBA fly ash is (7.8 ± 1.0) MPa, which is a 63 % decrease compared to the reference mixture without fly ash with a flexural strength of (21.0 ± 0.7) MPa. Significantly better flexural strengths were achieved by mixtures with a maximum of 30 wt.% FBA, which can be classified as Class AIII extruded tiles.
- A strong influence of quartz during cooling on Young's modulus. It can be expected that it is the formation of cracks during the modification of the quartz that results in a reduction of the flexural strength.
- A reduction in the shrinkage by firing at 30 and 40 wt.% FBA. This is not the case at 10 and 20 wt.% FBA, where the shrinkage by firing is slightly greater than for the reference mixture.
- A reduction in the bulk density. Again, this only applies to mixtures with 30 and 40 wt.% FBA.
- A decrease in the logarithmic decrement of the attenuation, especially after firing at temperatures below 1100 °C.
- Little effect on the thermal conductivity.
- FBA ash significantly deteriorated the properties of the ceramic bodies when it was present in the mixtures in quantities greater than 30 %. The main reason for the deterioration of the observed properties is the silica present in the FBA fly ash.

General observations are:

- The admixture of grog and fly ash makes the drying of illite-based ceramic bodies safer in terms of the risk of crack formation;

- During heating, volume changes up to ~800 °C are due to the behavior of illite, in particular, due to its dehydroxylation;
- Based on Young's modulus and logarithmic decrement of attenuation measurements, the maximum formation of microcracks during cooling (without considering the modification transformation of quartz) is at (250 – 350) °C, regardless of the presence of fly ash.

Several of these experimental results would require further study. For example, the high specific heat capacity values obtained by the comparative flash method, crystallization process of anorthite, evolution of Young's modulus in the sintering process, or phase composition. The isotropy of the materials properties was also assumed. However, the influence of the technological texture can significantly affect the directional dependence of the material parameters and, therefore, could have been a source of systematic errors.

In conclusion, based on the obtained results, it could be stated that power plant fly ash is a suitable substitute for traditional raw materials of building ceramics. In this monograph, a specific blend was not proposed that would immediately meet all requirements for the production of a specific building ceramics product from the forming process through drying, firing, and final utility properties. Possible applications include the production of bricks, wall and floor tiles, roof tiles, ceiling tiles, and other building ceramics products. The contribution of the study is in the number of analyses that have been carried out on experimental mixtures of illite and power plant fly ash to assess their correlation with each other. The results obtained can serve as a solid basis for developing a case study in which a natural source of clays from a deposit near a lignite-combusting thermal power plant is used instead of an illitic clay.

References

[1] I. Queralt, X. Querol, A. López-Soler, and F. Plana, "Use of coal fly ash for ceramics: a case study for a large Spanish power station," *Fuel*, vol. 76, no. 8, pp. 787–791, 1997. https://doi.org/10.1016/S0016-2361(97)00024-0

[2] F. Michalíková, J. Škvarla, and I. Brezáni, "Popolček ako prísada vo výrobe keramiky," ("Fly ash as an additive in the production of the ceramics") in *História a súčasný stav ťažby nerastných surovín na východnom Slovensku* (History and current state of mining of mineral resources in eastern Slovakia), 2011.

[3] F. Michalíková, Ľ. Floreková, and M. Benková, *Vlastnosti energetického odpadu - popola : využitie technológií pre environmentálne nakladanie* (Properties of energy waste - ash: use of technologies for environmental management), 1. Vydanie (1st edition). Košice: Technická Univerzita (Košice: Technical University), Slovakia, 2003.

[4] A. Mezencevová, "Možnosti zužitkovania energetických popolčekov," ("Energy fly ash utilisation options") *Acta Montan. Slovaca*, vol. 8, no. 2–3, pp. 146–151, 2003.

[5] S. Wang, C. Zhang, and J. Chen, "Utilization of Coal Fly Ash for the Production of Glass-ceramics With Unique Performances: A Brief Review," *J. Mater. Sci. Technol.*, vol. 30, no. 12, pp. 1208–1212, Dec. 2014. https://doi.org/10.1016/j.jmst.2014.10.005

[6] Z. Zhang, J. L. Provis, A. Reid, and H. Wang, "Fly ash-based geopolymers: The relationship between composition, pore structure and efflorescence," *Cem. Concr. Res.*, vol. 64, pp. 30–41, Oct. 2014. https://doi.org/10.1016/j.cemconres.2014.06.004

[7] R. S. Iyer, J. A. Scott, J. A. Au, and) Scott, "Power station fly ash-a review of value-added utilization outside of the construction industry," *Resour. Conserv. Recycl.*, vol. 31, pp. 217–228, 2001. https://doi.org/10.1016/S0921-3449(00)00084-7

[8] S. K. Mukherji, B. B. Machhoya, R. M. Savsani, D. R. Vyas, and T. K. Dan, "The utilisation of fly ash in the preparation of ceramic tableware and artware," *Br. Ceram. Trans.*, vol. 92, no. 6, pp. 254–257, 1993.

[9] N. Chandra, N. Agnihotri, S. Bhasin, and A. F. Khan, "Effect of addition of talc on the sintering characteristics of fly ash based ceramic tiles," *J. Eur. Ceram. Soc.*, vol. 25, no. 1, pp. 81–88, Jan. 2005. https://doi.org/10.1016/j.jeurceramsoc.2004.01.004

[10]　X. Lingling, G. Wei, W. Tao, and Y. Nanru, "Study on fired bricks with replacing clay by fly ash in high volume ratio," *Constr. Build. Mater.*, vol. 19, pp. 243–247, Apr. 2005. https://doi.org/10.1016/j.conbuildmat.2004.05.017

[11]　R. Sokolář, "Dry pressed ceramic tiles on the basis of fly ash," *InterCeram Int. Ceram. Rev.*, vol. 56, no. 1, pp. 30–35, 2007.

[12]　R. Sokolar and L. Smetanova, "Dry pressed ceramic tiles based on fly ash–clay body: Influence of fly ash granulometry and pentasodium triphosphate addition," *Ceram. Int.*, vol. 36, pp. 215–221, Jan. 2010. https://doi.org/10.1016/j.ceramint.2009.07.009

[13]　R. Sokolar and L. Vodova, "The effect of fluidized fly ash on the properties of dry pressed ceramic tiles based on fly ash–clay body," *Ceram. Int.*, vol. 37, pp. 2879–2885, Sep. 2011. https://doi.org/10.1016/j.ceramint.2011.05.005

[14]　A. Zimmer and C. P. Bergmann, "Fly ash of mineral coal as ceramic tiles raw material," *Waste Manag.*, vol. 27, no. 1, pp. 59–68, Jan. 2007. https://doi.org/10.1016/j.wasman.2006.01.009

[15]　N. Chandra, P. Sharma, G. L. Pashkov, et al., "Coal fly ash utilization: Low temperature sintering of wall tiles," *Waste Manag.*, vol. 28, pp. 1993–2002, 2008. https://doi.org/10.1016/j.wasman.2007.09.001

[16]　G. Cultrone and E. Sebastián, "Fly ash addition in clayey materials to improve the quality of solid bricks," *Constr. Build. Mater.*, vol. 23, no. 2, pp. 1178–1184, Feb. 2009. https://doi.org/10.1016/j.conbuildmat.2008.07.001

[17]　M. Krgović, M. Knežević, M. Ivanović, et al., "The properties of a sintered product based on electrofilter ash," *Mater. Tehnol. / Mater. Technol.*, vol. 43, no. 6, pp. 327–331, 2009.

[18]　S. Ferrari and A. F. Gualtieri, "The use of illitic clays in the production of stoneware tile ceramics," *Appl. Clay Sci.*, vol. 32, no. 1, pp. 73–81, 2006. https://doi.org/10.1016/j.clay.2005.10.001

[19]　J. K. Mitchell and K. Soga, *Fundamentals of Soil Behavior*, Third Edit. Hoboken, New Jersey: John Wiley & Sons, Inc., 2005.

[20]　Z. Pospíšil and A. Koller, *Fine ceramics. Introduction and foundations of technology (in Czech)*. Prague: Alfa, 1981.

[21]　J. Hlaváč, *Technology of Silicates*. Praha: SNTL, 1981.

[22]　P. Pytlík and R. Sokolář, *Building ceramics - technology, properties and applicataion*, (In Czech). Brno: CERM, 2002.

[23]　Z. Pánek, *Construction ceramics (in Slovak)*. Bratislava: R&D print, 1992.

[24] D. Antal, T. Húlan, A. Trník, I. Štubňa, and J. Ondruška, "The Influence of Texture and Firing on Thermal and Elastic Properties of Illite-Based Ceramics," *Adv. Mater. Res.*, vol. 1126, pp. 53–58, Oct. 2015. https://doi.org/10.4028/www.scientific.net/AMR.1126.53

[25] M. M. Erol, "Glass, glass-ceramics and sintered materials produced from industrial wastes," no. June, p. 327, 2006.

[26] L. Carbajal, F. Rubio-Marcos, M. A. Bengochea, and J. F. Fernandez, "Properties related phase evolution in porcelain ceramics," *J. Eur. Ceram. Soc.*, vol. 27, no. 13–15, pp. 4065–4069, 2007. https://doi.org/10.1016/j.jeurceramsoc.2007.02.096

[27] J. Wu, Z. Li, Y. Huang, and F. Li, "Crystallization behavior and properties of K2O-CaO-Al2O3-SiO2 glass-ceramics," *Ceram. Int.*, vol. 39, no. 7, pp. 7743–7750, 2013. https://doi.org/10.1016/j.ceramint.2013.03.031

[28] S. Sembiring, W. Simanjuntak, R. Situmeang, A. Riyanto, and K. Sebayang, "Preparation of refractory cordierite using amorphous rice husk silica for thermal insulation purposes," *Ceram. Int.*, vol. 42, no. 7, pp. 8431–8437, 2016. https://doi.org/10.1016/j.ceramint.2016.02.062

[29] O. Turkmen, A. Kucuk, and S. Akpinar, "Effect of wollastonite addition on sintering of hard porcelain," *Ceram. Int.*, vol. 41, no. 4, pp. 5505–5512, 2015. https://doi.org/10.1016/j.ceramint.2014.12.126

[30] T. Manfredini and M. Hanuskova, "Natural Raw Materials in 'Traditional' Ceramic Manufacturing," *J. Univ. Chem. Technol. Metall.*, vol. 47, no. 4, pp. 465–470, 2012.

[31] SII NanoTechnology Inc., "Thermal Analysis of Gypsum," *Application Brief. TA*, no. 2, Tokyo, pp. 1–3, Nov-1985.

[32] J. T. Kloprogge, H. D. Ruan, and R. L. Frost, "Thermal decomposition of bauxite minerals: Infrared emission spectroscopy of gibbsite, boehmite and diaspore," *J. Mater. Sci.*, vol. 37, no. 6, pp. 1121–1129, 2002.

[33] G. Hu, K. Dam-Johansen, S. Wedel, and J. P. Hansen, "Decomposition and oxidation of pyrite," *Prog. Energy Combust. Sci.*, vol. 32, no. 3, pp. 295–314, 2006. https://doi.org/10.1016/j.pecs.2005.11.004

[34] T. Húlan, A. Trník, and I. Medveď, "Kinetics of thermal expansion of illite-based ceramics in the dehydroxylation region during heating," *J. Therm. Anal. Calorim.*, vol. 127, no. 1, pp. 291–298, Jan. 2017. https://doi.org/10.1007/s10973-016-5873-0

[35] M. Khachani, A. El Hamidi, M. Halim, and S. Arsalane, "Non-isothermal kinetic and thermodynamic studies of the dehydroxylation process of synthetic calcium hydroxide Ca(OH)2," *J. Mater. Environ. Sci.*, vol. 5, no. 2, pp. 615–624, 2014.

[36] S. Gunasekaran and G. Anbalagan, "Thermal decomposition of natural dolomite," *Bull. Mater. Sci.*, vol. 30, no. 4, pp. 339–344, 2007. https://doi.org/10.1007/s12034-007-0056-z

[37] B. Kamphuis, A. W. Potma, W. Prins, and W. P. M. Vanswaaij, "The Reductive Decomposition of Calcium-Sulfate .1. Kinetics of the Apparent Solid Solid Reaction," *Chem. Eng. Sci.*, vol. 48, no. 1, pp. 105–116, 1993. https://doi.org/10.1016/0009-2509(93)80287-Z

[38] B. F. Bohor, "High-Temperature Phase Development in Illitic Clays," *Clays Clay Miner.*, vol. 12, no. 1, pp. 233–246, 1963. https://doi.org/10.1346/CCMN.1963.0120125

[39] M. Knapek, T. Húlan, P. Minárik, et al., "Study of microcracking in illite-based ceramics during firing," *J. Eur. Ceram. Soc.*, vol. 36, no. 1, pp. 221–226, 2016. https://doi.org/10.1016/j.jeurceramsoc.2015.09.004

[40] S.-J. L. Kang, *Sintering: Densification, Grain growth and Microstructure*. Oxford: Elsevier Butterworth-Heinemann, 2004.

[41] A. F. Gualtieri and S. Ferrari, "Kinetics of illite dehydroxylation," *Phys. Chem. Miner.*, vol. 33, no. 7, pp. 490–501, Oct. 2006. https://doi.org/10.1007/s00269-006-0092-z

[42] D. L. Carroll, T. F. Kemp, T. J. Bastow, and M. E. Smith, "Solid-state NMR characterisation of the thermal transformation of a Hungarian white illite," *Solid State Nucl. Magn. Reson.*, vol. 28, no. 1, pp. 31–43, 2005. https://doi.org/10.1016/j.ssnmr.2005.04.001

[43] "A Laboratory Manual for X-Ray Powder Diffraction - Illite group." [Online]. Available: https://pubs.usgs.gov/of/2001/of01-041/htmldocs/clays/illite.htm. [Accessed: 27-Jan-2020].

[44] R. E. Grim, R. H. Bray, and W. F. Bradley, "The Mica in Argillaceous Sediments†," *Am. Mineral.*, vol. 22, no. 7, pp. 813–829, 1937.

[45] J. Środoń and D. D. Eberl, "Micas," *Reviews in Mineralogy*, vol. 13, Mineralogical Society of America, Virginia, pp. 495–544, 1984. https://doi.org/10.1515/9781501508820-016

[46]   R. Ori, "The mineralogical and technological characterization of illite from Füzérradvány (Hungary) as a raw material for traditional ceramics," University of Modena and Emilia Region, Modena, 2003.

[47]   M. Rieder, G. Cavazzini, Y. S. D'yakonov, et al., "Nomenclature of the micas," *Mineralogical Magazine*, vol. 63, no. April, pp. 267–279, 1999. https://doi.org/10.1180/minmag.1999.063.2.13

[48]   P. Pytlík and R. Sokolář, *Building ceramics – technology, properties and application*. Brno: Akademické nakladatelství CERM, s.r.o., 2002.

[49]   A. F. Gualtieri, S. Ferrari, M. Leoni, et al., "Structural characterization of the clay mineral illite-1M," *J. Appl. Crystallogr.*, vol. 41, no. 2, pp. 402–415, 2008. https://doi.org/10.1107/S0021889808004202

[50]   G. H. Grathoff and D. M. Moore, "Illite Polytype Quantification using WILDFIRE© Calculated X-Ray Diffraction Patterns," *Clays Clay Miner.*, vol. 44, no. 6, pp. 835–842, 1996. https://doi.org/10.1346/CCMN.1996.0440615

[51]   N. Güven, "Comment on a Definition of Illite/Smectite Mixed-Layer," *Clays*, vol. 39, no. 6, pp. 661–662, 1991. https://doi.org/10.1346/CCMN.1991.0390613

[52]   H. E. Roberson and R. W. Lahann, "Smectite to Illite Conversion Rates: Effects of Solution Chemistry," 1981. https://doi.org/10.1346/CCMN.1981.0290207

[53]   J. Środoń, D. D. Eberl, and V. A. Drits, "Evolution of Fundamental-Particle Size during Illitization of Smectite and Implications for Reaction Mechanism," *Clays Clay Miner.*, vol. 48, no. 4, pp. 446–458, 2000. https://doi.org/10.1346/CCMN.2000.0480405

[54]   B. Kübler and M. Jaboyedoff, "Illite crystallinity," *Comptes Rendus l'Académie des Sci. - Ser. IIA - Earth Planet. Sci.*, vol. 331, no. 2, pp. 75–89, Jul. 2000. https://doi.org/10.1016/S1251-8050(00)01395-1

[55]   R. B. Furlong, "Electron Diffraction and Micrographic Study of the High-Temperature Changes in Illite and Montmorillonite Under Continuous Heating Conditions*," *Clays Clay Miner.*, vol. 15, no. 1, pp. 87–101, 1967. https://doi.org/10.1346/CCMN.1967.0150110

[56]   A. Aras, "The change of phase composition in kaolinite- and illite-rich clay-based ceramic bodies," *Appl. Clay Sci.*, vol. 24, no. 3–4, pp. 257–269, Feb. 2004. https://doi.org/10.1016/j.clay.2003.08.012

[57]   D. Wattanasiriwech, K. Srijan, and S. Wattanasiriwech, "Vitrification of illitic clay from Malaysia," *Appl. Clay Sci.*, vol. 43, no. 1, pp. 57–62, Jan. 2009. https://doi.org/10.1016/j.clay.2008.07.018

[58] J. Parras, C. Sánchez-Jimeńez, M. Rodas, and F. J. Luque, "Ceramic applications of Middle Ordovician shales from central Spain," *Appl. Clay Sci.*, vol. 11, no. 1, pp. 25–41, Oct. 1996. https://doi.org/10.1016/0169-1317(96)00003-8

[59] Z. Haiying, Z. Youcai, and Q. Jingyu, "Study on use of MSWI fly ash in ceramic tile," *J. Hazard. Mater.*, vol. 141, no. 1, pp. 106–114, Mar. 2007. https://doi.org/10.1016/j.jhazmat.2006.06.100

[60] M. Dondi, G. Ercolani, G. Guarini, and M. Raimondo, "Orimulsion fly ash in clay bricks—part 1," *J. Eur. Ceram. Soc.*, vol. 22, no. 11, pp. 1729–1735, Oct. 2002. https://doi.org/10.1016/S0955-2219(01)00493-9

[61] G. V. Rama Subbarao, D. Siddartha, T. Muralikrishna, K. S. Sailaja, and T. Sowmya, "Industrial Wastes in Soil Improvement," *ISRN Civ. Eng.*, vol. 2011, pp. 1–5, Sep. 2011. https://doi.org/10.5402/2011/138149

[62] S. Oka, *Fluidized Bed Combustion*. New York: Marcel Dekker Inc., 2004. https://doi.org/10.1201/9781420028454

[63] S. Hartuti, S. Kambara, A. Takeyama, K. Kumabe, and H. Moritomi, "Direct Quantitative Analysis of Arsenic in Coal Fly Ash," *J. Anal. Methods Chem.*, vol. 2012, no. 1, pp. 1–6, 2012. https://doi.org/10.1155/2012/438701

[64] C. Heidrich, H.-J. Feuerborn, and A. Weir, "Coal Combustion Products: a Global Perspective," in *2013 World of Coal Ash (WOCA) Conference*, 2013.

[65] B. H. Bowen and M. W. Irwin, "Coal Characteristics CCTR Basic Facts File # 8," 2008. [Online]. Available: http://www.purdue.edu/dp/energy/CCTR/. [Accessed: 31-Jan-2022].

[66] "Quarterly coal statistics," *International Energy Agency*, 2022. .

[67] D. Harris, C. Heidrich, and J. Feuerborn, "Global aspects on Coal Combustion Products.," *VGB PowerTech*, vol. 10, pp. 25–33, 2020.

[68] H.-J. Feuerborn, "Coal ash utilisation over the world and in Europe," in *Workshop on Environmental and Health Aspects of Coal Ash Utilization International workshop 23 rd-24 th*, 2005.

[69] U. Kleinhans, C. Wieland, F. J. Frandsen, and H. Spliethoff, "Ash formation and deposition in coal and biomass fired combustion systems: Progress and challenges in the field of ash particle sticking and rebound behavior," *Prog. Energy Combust. Sci.*, vol. 68, pp. 65–168, Sep. 2018. https://doi.org/10.1016/j.pecs.2018.02.001

[70] M. Ahmaruzzaman, "A review on the utilization of fly ash," *Prog. Energy Combust. Sci.*, vol. 36, no. 3, pp. 327–363, 2010. https://doi.org/10.1016/j.pecs.2009.11.003

[71] "EN450-1:2012 Fly ash for concrete - Part 1: Definitions, specifications and conformity criteria," 2012.

[72] "AS3582.1-1998 Supplementary cementitious materials for use with portland and blended cement - Fly ash," 1998.

[73] "ASTM C 618-03 Standard Specification for Coal Fly Ash and Raw or Calcined Natural Pozzolan for Use in Concrete," 2003.

[74] M. Erol, *Glass, glass-ceramic and sintered materials produced from industrial wastes,*. Istanbul, 2006.

[75] M. Erol, S. Küçükbayrak, and A. Ersoy-Meriçboyu, "Comparison of the properties of glass, glass–ceramic and ceramic materials produced from coal fly ash," *J. Hazard. Mater.*, vol. 153, pp. 418–425, May 2008. https://doi.org/10.1016/j.jhazmat.2007.08.071

[76] E. ul Haq, S. Kunjalukkal Padmanabhan, and A. Licciulli, "Synthesis and characteristics of fly ash and bottom ash based geopolymers–A comparative study," *Ceram. Int.*, vol. 40, no. 2, pp. 2965–2971, Mar. 2014. https://doi.org/10.1016/j.ceramint.2013.10.012

[77] F. Škvára, N. A. Duong, and Z. Zlámalová Cílová, "Geopolymer materials on the flyash basis - Long-term properties," *Ceram. - Silikaty*, vol. 58, no. 1, pp. 12–20, 2014.

[78] R. S. Iyer and J. A. Scott, "Power station fly ash - A review of value-added utilization outside of the construction industry," *Resour. Conserv. Recycl.*, vol. 31, no. 3, pp. 217–228, 2001. https://doi.org/10.1016/S0921-3449(00)00084-7

[79] M. Visa, "Synthesis and characterization of new zeolite materials obtained from fly ash for heavy metals removal in advanced wastewater treatment," *Powder Technol.*, vol. 294, pp. 338–347, Jun. 2016. https://doi.org/10.1016/j.powtec.2016.02.019

[80] I. Acar and M. U. Atalay, "Characterization of sintered class F fly ashes," *Fuel*, vol. 106, pp. 195–203, 2013. https://doi.org/10.1016/j.fuel.2012.10.057

[81] M. Erol, S. Küçükbayrak, and A. Ersoy-Meriçboyu, "Characterization of sintered coal fly ashes," *Fuel*, vol. 87, pp. 1334–1340, Jun. 2008. https://doi.org/10.1016/j.fuel.2007.07.002

[82] O. Kayali, "High Performance Bricks from Fly Ash," *2005 World Coal Ash*, pp. 1–13, 2005.

[83] A. Olgun, Y. Erdogan, Y. Ayhan, and B. Zeybek, "Development of ceramic tiles from coal fly ash and tincal ore waste," *Ceram. Int.*, vol. 31, pp. 153–158, Jan. 2005. https://doi.org/10.1016/j.ceramint.2004.04.007

[84] T. Húlan, I. Medveď, A. Trník, and R. Podoba, "Kinetics of high-temperature sintering in an illite-based ceramic body studied by thermodilatometry," in *Thermophysics 2014 - Conference Proceedings, 19th International Meeting of Thermophysical Society*, 2014, pp. 53–60.

[85] T. Húlan, A. Trník, I. Štubňa, et al., "Development of Young's Modulus of Illitic Clay during Heating up to 1100 °C," *Mater. Sci.*, vol. 21, no. 3, pp. 429–434, Sep. 2015.

[86] G. Crolly, "Fritsch Analysette 22 Laser particle Sizer," 2009.

[87] S. Brunauer, P. Emmett, and E. Teller, "Adsorption of gases in multimolecular layers," *J. Am. Chem. Soc.*, vol. 60, no. 2, pp. 309–319, 1938. https://doi.org/10.1021/ja01269a023

[88] M. Földvári, *Handbook of thermogravimetric system of minerals and its use in geological practice*. Budapest, 2011.

[89] R. Podoba, A. Trník, and Ľ. Podobník, "Upgrading of TGA/DTA analyzer Derivatograph," *Építőanyag*, vol. 64, no. 1–2, pp. 28–29, 2012. https://doi.org/10.14382/epitoanyag-jsbcm.2012.5

[90] I. Štubňa, A. Trník, and L. Vozár, "Thermomechanical and thermodilatometric analysis of green alumina porcelain," *Ceram. Int.*, vol. 35, no. 3, pp. 1181–1185, Apr. 2009. https://doi.org/10.1016/j.ceramint.2008.05.004

[91] I. Štubňa, A. Vážanová, G. Varga, and D. Hrubý, "Simple push-rod dilatometer for dilatometry of ceramics," in *Reserach and Teaching of Physics in the Context of University Education*, 2007, pp. 69–74.

[92] C.-L. Hwang and T.-P. Huynh, "Investigation into the use of unground rice husk ash to produce eco-friendly construction bricks," *Constr. Build. Mater.*, vol. 93, no. 2015, pp. 335–341, Sep. 2015. https://doi.org/10.1016/j.conbuildmat.2015.04.061

[93] P. Martinez, M. Soto, F. Zunino, C. Stuckrath, and M. Lopez, "Effectiveness of tetra-ethyl-ortho-silicate (TEOS) consolidation of fired-clay bricks manufactured with different calcination temperatures," *Constr. Build. Mater.*, vol. 106, pp. 209–217, Mar. 2016. https://doi.org/10.1016/j.conbuildmat.2015.12.116

[94] "ASTM C 1161 - 02c, ASTM C 1161 - 02c. Standard test method for flexural strength of advanced ceramics at ambient temperature," 2002.

[95] V. Mikli, H. Käerdi, P. Kulu, and M. Besterci, "Characterization of powder particle morphology," *Proc. Est. Acad. Sci. Eng.*, vol. 7, no. 1, pp. 22–34, 2001. https://doi.org/10.3176/eng.2001.1.03

[96] F. Michalíková, *Vlastnosti energetického odpadu – popolčeka* (Properties of energy waste - fly ash). Košice: Krivda, Slovakia, 2003.

[97] S. Guggenheim, Y.-H. Chang, F. Koster Van Groos, A., Y.-H. Vhang, and F. Koster Van Groos, A., "Muscovite dehydroxylation: High-temperature studies," *Am. Mineral.*, vol. 72, pp. 537–550, 1987.

[98] V. A. Drits, H. Lindgreen, A. L. Salyn, R. Ylagan, and D. K. McCarty, "Semiquantitative determination of trans-vacant and cis-vacant 2:1 layers in illites and illite-smectites by thermal analysis and X-ray diffraction," *Am. Mineral.*, vol. 83, no. 11-12 PART 1, pp. 1188–1198, 1998. https://doi.org/10.2138/am-1998-11-1207

[99] Q. L. Yu and H. J. H. Brouwers, "Thermal properties and microstructure of gypsum board and its dehydration products: a theoretical and experimental investigation," *Fire Mater.*, vol. 36, no. 7, pp. 575–589, 2012. https://doi.org/10.1002/fam.1117

[100] V. Tydlitát, A. Trník, L. Scheinherrová, R. Podoba, and R. Černý, "Application of isothermal calorimetry and thermal analysis for the investigation of calcined gypsum-lime-metakaolin-water system," *J. Therm. Anal. Calorim.*, vol. 122, no. 1, pp. 115–122, 2015. https://doi.org/10.1007/s10973-015-4727-5

[101] T. Húlan, A. Trník, T. Kaljuvee, et al., "The study of firing of a ceramic body made from illite and fluidized bed combustion fly ash," *J. Therm. Anal. Calorim.*, vol. 127, no. 1, pp. 79–89, Jan. 2017. https://doi.org/10.1007/s10973-016-5477-8

[102] A. Trník, I. Štubňa, R. Sokolář, and I. Medveď, "Use of fly ash in ceramic tiles: elastic properties during firing," *J. Ceram. Soc. Japan*, vol. 121, no. 1419, pp. 925–929, 2013. https://doi.org/10.2109/jcersj2.121.925

[103] F. Chmelík, A. Trník, I. Štubňa, and J. Pešička, "Creation of microcracks in porcelain during firing," *J. Eur. Ceram. Soc.*, vol. 31, no. 13, pp. 2205–2209, 2011. https://doi.org/10.1016/j.jeurceramsoc.2011.05.045

[104] L. Svoboda, M. Bažantová, M. Myška, et al., *Stavebné materiály* (Building materials), ASB. Bratislava: Jaga group, s.r.o., Slovakia, 2005.

[105] M. Jankula, T. Húlan, I. Štubňa, et al., "The influence of heat on elastic properties of illitic clay Radobica," *J. Ceram. Soc. Japan*, vol. 123, no. 1441, pp. 874–879, 2015. https://doi.org/10.2109/jcersj2.123.874

[106] I. Štubňa, P. Šín, A. Trník, R. Podoba, and L. Vozár, "Development of Young's modulus of the green alumina porcelain raw mixture," *J. Aust. Ceram. Soc.*, vol. 50, no. 2, pp. 36–42, 2014.

[107] I. Štubňa, M. Mánik, T. Húlan, and A. Trník, "Development of stress on quartz grain in illite ceramics during cooling stage of firing," *J. Ceram. Soc. Japan*, vol. 128, no. 3, pp. 117–123, Mar. 2020. https://doi.org/10.2109/jcersj2.19169

[108] M. S. Blantner, I. S. Golovin, H. Neuhäuser, and H.-R. Sinning, *Internal Friction in Metalic Materials*. Berlin: Springer-Verlag Berlin Heidelberg, 2007. https://doi.org/10.1007/978-3-540-68758-0

[109] "EN 14411 Ceramic tiles - Definitions, classification, characteristics and marking," 2004.

[110] Z. Weiss and A. Wiewióra, "Polytypism of Micas. III. X-Ray Diffraction Identification," *Clays Clay Miner.*, vol. 34, no. 1, pp. 53–68, 1986. https://doi.org/10.1346/CCMN.1986.0340107

About the Authors

Dr. Húlan is working as an associate professor at the Department of Physics, Constantine the Philosopher University in Nitra, Slovakia. At the same institution, he graduated in physics of materials and also completed his Ph.D. (2016). He has been working in the field of clay-based ceramic materials with a focus on the processes occurring during thermal treatment, including their kinetics, microcracking, and phase transitions. He has been involved in funded projects dedicated to the reuse of waste materials, the development of anorthite ceramics, and glass-ceramic glazes in the capacity of a principle and co-investigator. Dr. Húlan published over 40 journal articles and conference publications that received over 140 citations to these publications.

Dr. Ondruška is working as an associate professor at the Department of Physics, Constantine the Philosopher University in Nitra, Slovakia. He graduated in the teaching of physics and informatics (2007), and completed his Ph.D. (2011) at the same institution. The main focus of scientific activity is in condensed matter physics and acoustics, specializing in mechanical, electrical, and thermophysical properties of materials. He specializes in the study of the DC and AC electrical conductivity of ceramic materials in the temperature range of 25 – 1200 °C and the study of the mechanical and thermophysical properties of ceramic and building materials.

www.ingramcontent.com/pod-product-compliance
Lightning Source LLC
Chambersburg PA
CBHW071700210326
41597CB00017B/2268